AIR BAGS

. . . For the safety of occupants, it has become clear that autos need a passive protection system.

"Passive" means that the system must require no effort by the passengers,
and "protection" means that they must be safe during the equivalent of a frontal collision with a stationary barrier at 30 miles per hour. . . .

The electro-mechanical system consists of six main components, the most important being
a sensor. Slightly larger than a golf ball, the sensor is mounted on the automobile's
firewall. Its function is to determine when the auto has been in a collision. A spring
and a metal weight are located next to the sensor and combine with it to form an electrical circuit. Mounted behind the instrument panel is a bottle of nitrogen gas which has
a high-explosive cap. A tube connects the bottle to a coated-nylon bag folded up in
the instrument panel. . . .

As diagrammed in Figure 1, the air bag activates when an impact forces the sensor against
the spring. If the impact equals a barrier crash of 8 miles per hour, the spring moves the
metal weight approximately 1/2 inch, completing the electrical circuit. Electric wires
carry this signal to the nitrogen bottle's high-explosive cap. When the cap explodes,
the nitrogen rushes out into the pipe and is quickly distributed to the air bag. . . .

The need for a passive protection system in automobiles is shown in Table 1. Despite the
fact that safety belts are now standard equipment, over 50,000 Americans are killed each
year in auto accidents, and another 2 million are injured. These numbers could be reduced if safety equipment were worn, but the statistics show that few people wear it. Air
bags could save at least 40 percent of the lives now lost in frontal collisions. . . .

TABLE 1   DATA ON AUTO OCCUPANTS

Occupants using shoulder harnesses	5%
Occupants using seat belts	30%
Injuries per year	2 million
Deaths per year	54,000

According to Robert Lund, the system activates, inflates the air bag, and deflates the
bag in less than a second from the time of impact.[1] . . .

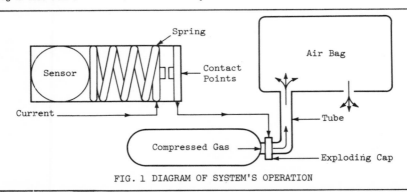

FIG. 1 DIAGRAM OF SYSTEM'S OPERATION

Technical Report Writing Today

Technical Report Writing Today

Second Edition

Steven E. Pauley
Purdue University

Houghton Mifflin Company   Boston

Dallas   Geneva, Illinois   Hopewell, New Jersey   Palo Alto   London

For my father, Orrin F. Pauley

Printed in the U.S.A.
Library of Congress Catalog Card Number: 78-69557
ISBN: 0-395-27111-8

Contents

Preface

The organization, approach, and objectives of the second edition of *Technical Report Writing Today* are the same as for the first: The book is designed to bridge the gap between the writing specialist who teaches the course and technical specialists from widely varied disciplines who need to improve their communication skills. Student and instructor meet each other halfway. Students write papers on topics from their specialized fields and aim their writing at an uninformed reader—the instructor.

This approach helps students because it reduces the unknowns. If the information is already known to students, they can concentrate less on what to say and more on how to say it well (the unknown). Having been given the freedom to select their own topics, however, students assume the responsibility of communicating with an instructor who will understand the information only if it is clearly written and includes explanations of specialized terminology. This kind of writing is not just an academic exercise. After graduation, students will not be asked to write reports telling a supervisor what he or she already knows.

The second edition of *Technical Report Writing Today* incorporates many suggestions from instructors who used the first edition. New examples and explanations illustrate the specific parts of a report. At the ends of chapters are new student models that are understandable by people with limited technical backgrounds, useful for classroom lecture and discussion, and helpful in demonstrating that the writing assignments are applicable to all fields. The text also includes major additions to the chapters on illustrating, operation manuals, letters of application, and business letters.

All but one of the student models in this book were written by freshmen and sophomores. The models are imperfect, but their organization gives students a pattern to follow. Students should be encouraged to examine the models critically and to identify their strengths and weaknesses. As they debate the accuracy and clarity of the models, students will become more aware of the need for those qualities in their own reports. The instructor should ask students to provide specific suggestions for improving the models and then require them to apply those suggestions in their own writing.

Many of the models contain visuals. Students should be cautioned not to expect visuals to do their communicating for them. The function of a visual is to reinforce words, not to substitute for good writing. Students who have not studied drawing should not be concerned about their lack of skill; a two-dimensional sketch can communicate more information than an ineffectively used three-dimensional drawing.

This book lends itself to use in both basic and advanced writing courses. Freshmen and sophomores generally devote most of the term to becoming skilled in the basic techniques of technical writing. For a final paper, they apply the techniques they have learned to a formal report, a research paper, an assignment that integrates both report and research, or a manual. Juniors and seniors generally concentrate on letters and formal reports. For their formal reports, they define problems in their major fields and then attempt to solve those problems in reports aimed at an uninformed reader.

I wish to thank the following students for giving me permission to print their reports:

Tom Clawson: "Sonar"
Steve Medanic: "Turbocharger"
Elmer Paulsen: "Description of a Soldering Iron"
Mark Tomko: "Description of a Bow Compass"
Jane Dechnik: "Description of an Insulin Syringe"
Tom Yokovich: "Description of an Incandescent Lamp"
Cameron Perryman: "Description of the Soldering Process"
Robert Dekker: "Description of a Thermostat in Operation"
Richard Kinney: "Developing a Roll of Film"
Ron Pawlik: "Recommendation of a Water Heater"
Richard Kinney: "Recommendation of a Voltmeter"
Clifford Spryka: "Recommendation of a Truck Tractor"
Janis Taylor: "Computers That Listen"
Gregory A. Stutz: "Solar Energy for Homes"
Jim Biggs: "Memorandum: Kauzlarich Construction Company"
Clinton Hare: "Memorandum: Garlitz Cement Company"
Thomas E. McKain: "Memorandum: N. P. Scott Manufacturing Company"
Paul Markovich: "Proposal for Protection Against Spread of Alpha Radioactivity"
Robert J. Smith: "Proposal for Installation of a Power Generator"
William A. Smolen: "Feasibility Report: Leasing or Purchasing a Crane"
Laurie Atzhorn: "The Compound Microscope"
Beth Lakin: "Operation Manual: Granada 23-channel Citizens' Band Transceiver"

I am grateful to the following persons for reading the manuscript and offering helpful suggestions: David H. Karrfalt, Edinboro State College; Emil A. Mucchetti, Texas A & I University; and Rodelle Weintraub, Pennsylvania State University.

I also want to thank my friends and colleagues for their patience during my work on this second edition.

S. E. P.

Section One

Introduction

1/THE TECHNICAL WRITER AND THE REPORT READER

The introductory chapter of this text analyzes the technical writer and the reader of technical reports. Most technical writing problems result from people aiming their reports at themselves rather than at others. In your role as a technical writer, you take the first step toward your overall objective, communication, when you fully understand the relationship between yourself and your reader.

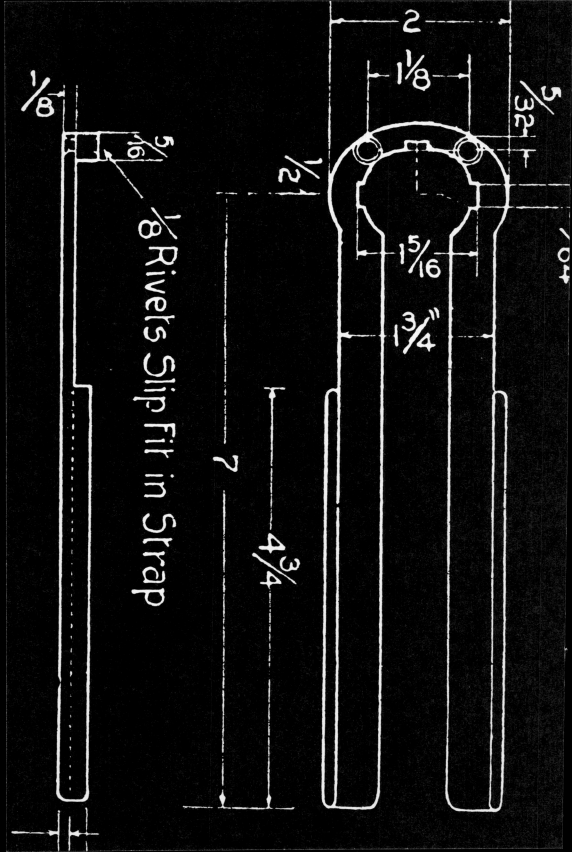

⅛ Rivets Slip Fit in Strap

1 The Technical Writer and the Report Reader

A Report-oriented Society

"Put it in writing."

These words are heard every day in industry. Both industry and society in general have become report oriented: not much counts or becomes official until it is on paper.

Such an emphasis on reports seems very impersonal, but it is the inevitable result of industry's growth. Decision-makers today just do not have 25 minutes to spend listening to an explanation of how something works, why it went wrong, or how it can be improved. They want the facts in a report that is as concise and readable as possible. Therefore, technologists spend up to 25% of their time writing up what they have been working on the other 75% of the time.

Anyone in industry knows the impact of that 25%. A technical report, on reaching the reader's desk, had better speak well for the person who wrote it. Promotions obviously depend on performance, and report writing is a highly visible and persuasive part of performance.

Although your technical knowledge will win that first job in industry, your report-writing ability will be important to your advancement. To put it bluntly, when you and another person with equal technical ability join the same firm, the one who can communicate better will advance first.

As your career progresses, report-writing skill will become increasingly valuable to you, for positions of authority carry greater writing responsibilities. Promotion to more desirable positions will become more and more dependent on your ability to present yourself and your ideas in writing.

Becoming a Better Writer

No one questions the need for better report writing in business and industry. Journals and employers in practically all technical fields lament the failure of technicians and engineers to communicate effectively. The fact that technical writing courses are now mandatory in technical curriculums reflects that need.

Despite the importance of writing, many technical students deny themselves the opportunity to become better writers. To realize their writing potential, they must reject the following rationalizations, which so often accompany failure.

A Lot of Great Scientists Are Lousy Writers. This rationalization is based on the mistaken impression that a scientist who comes up with a great idea automatically becomes known as a great scientist. In reality, the great scientist is one who comes up with a great idea and communicates it well. Albert Einstein had to spell out his complex, revolutionary ideas, and we need only glance at his simple writing style to know that he did a good job of it. Had Einstein been a poor communicator, people today might say, "Albert who?"

This Stuff Is Too Technical to Explain in Simple Words. Translated, this statement means either "I don't understand this stuff well enough myself to explain it to you," or "I don't want to make the effort to explain it to you." People who are proficient in their fields have the ability to communicate their information to those with less knowledge.

Louis Rukeyser, host of one of public television's highest-rated programs, "Wall Street Week," gives two reasons for his success in communicating complex information about the economy and stock market to less-informed people:

> First, you have to speak English. You can't hide behind jargon. You've got to talk about complicated things in a straightforward fashion. Second, if you do speak English, a lot of people will understand what you are saying. So you have to know what you're talking about.[1]

Industry rewards technologists who have precisely these two strengths: knowledge of what they are talking about, and the ability to communicate their knowledge in straightforward technical reports. If you thoroughly understand your subject and make the effort to apply the basic techniques of technical writing, you can write successful reports.

I'm Good at Technical Subjects, but English Loses Me. The saddest thing about this rationalization is that if you are convinced you can never improve your writing, your writing will never improve; time and instruction can do very little. For whatever it is worth, however, most teachers find that students who do well at technical subjects also do well at technical writing. Conversely, those who have difficulty with technical writing often have problems

---

[1] Louis Rukeyser as quoted by Bill Granger in *The Chicago Sun-Times*, April 28, 1978. Reprinted with permission from The Chicago Sun-Times.

in other technical courses. Exceptions do exist, of course, and anyone who chooses to rationalize is free to do so. If you give this course the same time you give your other subjects, however, you are likely to find that technical writing offers opportunities that perhaps were not available in your previous writing courses.

In a technical writing course, your writing topics are generally taken from your technical field or general technical background. This enables you to write from the strength of knowledge and to concentrate more on how to say it well than on what to say. You might also be more comfortable with the technical writing style than you were with the style prescribed in other composition courses. Technical writing emphasizes the concise, objective presentation of factual data. Therefore, reports seldom contain the first-person *I* or the second-person *you*; instead, they are written in the third-person *it,* which emphasizes the impartial recording of information such as physical characteristics, operational characteristics, and statistical evidence.

Writing Helps Thinking

Why, during the last paragraph of an in-class theme, with 3 minutes to go, do you suddenly figure out what you should have been saying all along? Why does writing a rough draft bring you closer to what you want to say? The writing process helps bring thoughts into focus. The words used for writing are the same ones used for thinking, and once you have written them, you can see them clearly and organize your rewriting.

The old definition of a sentence as a complete thought is still accurate. Many people like to blame poor writing on grammar and punctuation—which are easy, defenseless targets—instead of on poor thinking. Bad writing, however, is often but a symptom of inadequate thinking; if your thoughts are fragmented, indiscriminately spliced, underdeveloped, and disorganized, your writing will be the same.

The solution is to write rough drafts of reports, using the writing process to gain greater command of your thoughts. As you become more experienced at formulating and expressing these thoughts, you will need to write fewer rough drafts. Most technical writing students find that they become more knowledgeable about the writing topics they choose. Because writing helps thinking, they emerge from the course with a better understanding of their technical fields.

The Uninformed Reader

In the world of business and industry, you will generally write for an uninformed reader. If that seems odd, think of it this way: If the reader knew more about the subject than you did, he or she would be the writer and you would be the reader.

When asked to write a report for industry, you go through a long process of investigation, thinking, and writing. On completing this process, you know your subject forward and backward, and when you read the report to yourself, everything seems quite clear. But then you turn it over to readers who have not gone through the investigating, thinking, and writing process you have. The readers get it cold. Compared with you, they are uninformed about the subject. This does not mean that they are stupid or that things must be repeated for them. It does mean that you must write the report clearly if you are to communicate.

Achieving Communication

One key to good report writing is using words to communicate with people rather than to try to impress them. Anybody can sound impressive without making good sense. Phillip Broughton makes this frighteningly clear with his Systematic Buzz-phrase Projector (Table 1.1). Select any three-digit number,

Table 1.1   Systematic buzz-phrase projector

Column 1	Column 2	Column 3
0. Integrated	0. Management	0. Options
1. Total	1. Organizational	1. Flexibility
2. Systematized	2. Monitored	2. Capability
3. Parallel	3. Reciprocal	3. Mobility
4. Functional	4. Digital	4. Programming
5. Responsive	5. Logistical	5. Concept
6. Optional	6. Transitional	6. Time-phase
7. Synchronized	7. Incremental	7. Projection
8. Compatible	8. Third-generation	8. Hardware
9. Balanced	9. Policy	9. Contingency

Philip S. Broughton, "Criteria for the Evaluation of Printed Matter," *American Journal of Public Health,* 30 (September 1940), 1027–1032.

find the word in each column that corresponds to each digit of the number, and you will have a tremendously impressive phrase. The number 757, for example, produces "synchronized logistical projection," a phrase that could be inserted into any report to impress the reader. As Broughton says, nobody will know what it means but nobody will admit it, either. Industry has no use for this kind of writing. The only really impressive writing is writing that communicates.

Gunning's Fog Index

Robert Gunning created the Fog Index to serve as a practical yardstick for determining the degree of difficulty in any type of writing. It works like this:

> *One:* Jot down the number of words in successive sentences. If the piece is long, you may wish to take several samples of 100 words, spaced evenly through it. If you do, stop the sentence count with the sentence which ends nearest the 100-word total. Divide the total number of words in the passage by the number of sentences. This gives the average sentence length of the passage.
>
> *Two:* Count the number of words of three syllables or more per 100 words. Don't count the words (1) that are proper names, (2) that are combinations of short easy words (like "bookkeeper" and "manpower"), (3) that are verb forms made three syllables by adding -*ed* or -*es* (like "created" or trespasses"). This gives you the percentage of hard words in the passage.
>
> *Three:* To get the Fog Index, total the two factors just counted and multiply by .4.[2]

Using a passage written by Albert Einstein, Gunning demonstrates that complex ideas can be expressed in clear, fog-free writing. Let's apply the Fog Index to the passage:

> If we ponder over the question as to how the *universe, considered* as a whole, is to be regarded, the first answer that suggests itself to us is surely

[2] Robert Gunning, *The Technique of Clear Writing,* rev. ed. (New York: McGraw-Hill, 1968), p. 38.

this: As regards space (and time) the *universe* is *infinite*. There are stars everywhere, so that the *density* of matter, although very *variable* in detail, is nevertheless on the *average* everywhere the same. In other words: However far we might travel through space, we should find everywhere an *attenuated* swarm of fixed stars of *approximately* the same kind of *density*.[3]

Total words: 89
Total sentences (independent clauses count as sentences): 4
Average sentence length: $89 \div 4 = 22.2$
Total three-syllable-or-more words: 10
Percentage of three-syllable-or-more words: $10 \div 89 = 11.2$

$22.2 + 11.2 = 33.4$
$33.4 \times 0.4 = 13.3$ *Fog Index*

To determine what the 13.3 means, let's again turn to Gunning, who compares the Fog Index with the reading levels required for various magazines and grades in school (Table 1.2). The table indicates the reading levels of popular

Table 1.2   Reading level

	Fog index	By grade		By magazine
	17	College graduate		
	16	"	senior	(No popular magazine this
	15	"	junior	difficult.)
Danger	14	"	sophomore	
Line	13	"	freshman	
	12	High-school senior		*Atlantic Monthly* and *Harper's*
	11	"	junior	*Time* and *Newsweek*
	10	"	sophomore	*Reader's Digest*
	9	"	freshman	*Saturday Evening Post*
Easy-	8	Eighth grade		*Ladies' Home Journal*
reading	7	Seventh "		*True Confessions* and *Modern*
Range				*Romances*
	6	Sixth "		Comics

Robert Gunning, *The Technique of Clear Writing*, rev. ed. (New York: McGraw-Hill, 1968), p. 40. Reprinted with permission.

[3] Albert Einstein, quoted in Gunning, *Clear Writing*, p. 256.

magazines. They seem to be aimed very low, but of course their circulation depends on reaching uninformed readers. The only exception to Gunning's statement that no popular magazine indexes above 12 is *Playboy,* which might be evidence that more people look at it than read it. According to the table, Einstein's passage could be understood by a college freshman, which is amazing in view of the subject matter. One of Einstein's theories must have been "Keep writing simple in order to communicate."

Gunning's Fog Index emphasizes that reports should not make information more difficult to understand than it already is. The ways to present technical information clearly are to use understandable rather than needlessly difficult terminology, and to use shorter sentences when you explain particularly complex information. According to a nationwide study by Richie R. Ward,[4] even highly educated and experienced technologists want reports written below the Fog Index of 15; to make the right decisions based on information in reports, readers need to understand the information thoroughly. Some writers make the mistake of thinking that because the material is complex, it cannot be communicated clearly; industry's response is that because the material is complex, it *must* be communicated clearly.

The Value of Visuals

Visuals can greatly assist your effort to communicate technical information, for most people are more adept at grasping visual information than information presented in words. Technical visuals, including photographs, drawings, charts, diagrams, and tables, make reports more attractive; more important, they reinforce the paragraphs of a report. You can select the type of visual that corresponds to your subject matter and to your reader's level of technical knowledge. Chapter 7 provides a detailed explanation of how to use visuals effectively.

Too many reports are written in a vacuum: writers write for themselves instead of aiming at uninformed readers, and then they wonder why their reports fail to communicate. Become conscious of your readers, and try to write reports that defy misunderstanding. This kind of fail-safe writing, from your readers' point of view, is the only good technical writing.

[4] *Practical Technical Writing* (New York: Knopf, 1968).

Section Two
Technical Writing Techniques

This section of the text describes the basic techniques that you will need for writing for industry. Practically every word in your reports will fall into one of these categories: definition, description, or interpretation of statistical data. Your descriptions will not always be limited to mechanisms, of course, but the principles for describing mechanisms and their operations will apply to your reports in industry. Research techniques will prepare you for other academic assignments as well as for keeping aware of advancements in your field after you graduate. Illustrating is not specifically a writing technique, but it does enable you to reinforce the words in your reports.

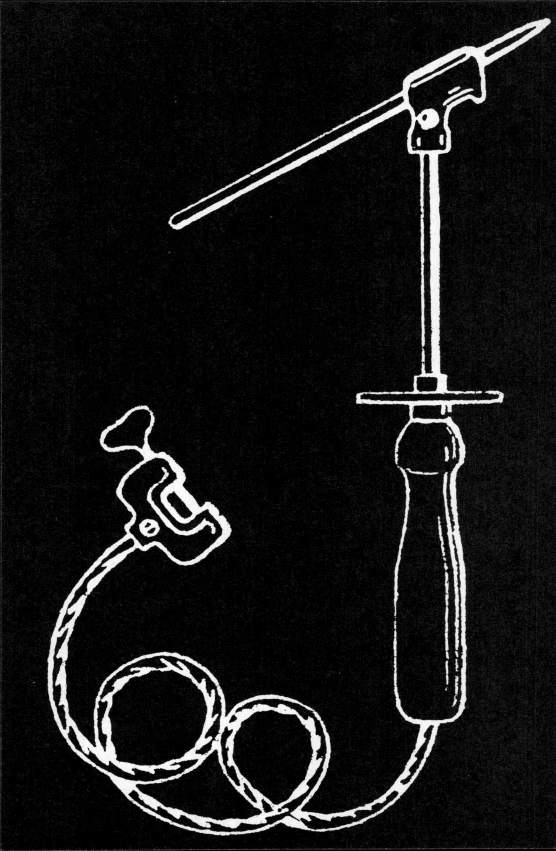

2 Defining

DEFINITION is a logical technique that reveals the meaning of a term. It has become important in this era of specialization because each field has developed words that are not familiar to people outside that field. Very few automotive technologists, for example, can wander into an electronics laboratory, spend 5 minutes listening to words like *heterodyne detection, pyrometer,* and *micromho,* and emerge with more than a vague idea of what is going on. A specialist must recognize that the field *is* special, that it has specialized terminology unknown to people in other fields, and that defining terms is a prerequisite to clear communication.

Specialists working in industry or in the nonindustrial world of institutions (foundations, universities, medical centers) and governmental agencies (local, state, federal) must be adept at defining terms. Specialized terminology is appropriate for informal reports within a department, but not for reports and memorandums going to other departments. For example, a company attempting to develop a synthetic motor oil must encourage communication among specialists in chemical technology, automotive technology, pollution control, and metallurgy. Such a project will not succeed unless technical specialists write clear reports to each other stating their ideas and their test results.

Even if communication among various technologies presented no difficulty, the problem of communication with management would still exist. Not only has the number of technologies multiplied, but management has become diversified and specialists are required to administer particular areas. Although some management personnel have come up through the ranks and others have strong technical backgrounds, they cannot realistically be expected to keep up with various technologies while they function as administrators.

Nevertheless, technical specialists must communicate with management. The company's administrators will allocate a great deal of money to, say, the synthetic oil project, and they will expect regular progress reports. If the metallurgists write confusing reports, administrators reading them will be justified in concluding that the metallurgists are confused. If the automotive specialists want more money for their portion of the project, they must explain their needs to the people who control the money. If the technologists cannot explain their new product to the marketing department, how can the marketing people convince the public to buy it? In any company, people with very different areas of expertise must communicate with each other in a coordinated effort to meet the company's objectives.

In short, the era of generalists—persons with working knowledge of several fields—has passed. People have enough trouble keeping up with innovations in their own fields. Report writers today recognize this and write accordingly. Words are carefully chosen and defined because no reader should have to ask for clarification. Neither should a reader be allowed to think she or he knows what a word means, only to discover later that the intended meaning was entirely different. Reports are written to save time rather than waste it.

As Chapter 1 emphasized, communicating information to uninformed readers is part of every technician's job. This chapter concentrates on defining, one of the most important techniques for doing that job successfully.

Informal Definitions

Avoid definitions whenever possible; if a simple word conveys the intended meaning, use it. However, when you cannot avoid technical terminology, an informal definition is often sufficient. Though less thorough than the formal sentence definitions explained later in this chapter, the following kinds of informal definitions are quite acceptable if they furnish the necessary information.

OPERATIONAL DEFINITIONS

Operational definitions are partial definitions that explain what something does. For example, an informal, operational definition of a *centrifuge* would state that it separates materials with different densities. *Acceleration* can be partially defined by an explanation of what happens when an automobile driver steps on the gas pedal. An operational definition of *frictional electricity* might state that it negatively charges a person as a result of contact between the person's shoes and a rug, and that it causes a spark and a small shock when he or she touches an object and the excess electrons rush off.

NEGATIVE STATEMENTS

Negative statements explain what a term does not mean, often in order to correct a common misconception. To be effective, a negative statement must be followed by a positive one. You might, for instance, say that a *funnycar* is not necessarily a funny-looking car. You would then explain that *funny* derives from *phoney,* which refers to the car's having a fiber-glass rather than a metal body.

A series of negative definitions can be used in a process of elimination leading to the exact definition. You might explain that a construction technologist cannot properly be called either a mechanical artist or an architect, and then make distinctions among the three to arrive at an accurate, positive definition.

SYNONYMS

A synonym is effective as a definition only when it is better known than the term being defined. People are more familiar with *voltage* than *electromotive force,* the technical term carrying the same meaning. Other examples are *loudspeaker* for *electroacoustic transducer* and *spun glass* for *fiber glass.* When aiming at an uninformed reader, never pass up an opportunity to clarify a technical term with a more common word; if possible, avoid the difficult word entirely by merely substituting the easier one.

In daily conversation, people begin informal definitions by saying "What I mean is . . .," or "Look at it this way." You can use the same conversational approach to clarify words and ideas in technical reports. Nothing is wrong with writing "In other words . . ." or "In this report, *environment* refers to . . .," or with simply placing a synonym in parentheses. Clear and conversational writing communicates with readers because it presents technical information in their own language.

Formal Definitions

Every technology has its precision instruments. Learning how to use them requires patience, but they ultimately make the job easier. In technical writing, the formal sentence definition functions like one of those instruments. It separates the item being discussed from everything else in the world. The definition's three parts are the item (species) that needs defining, the class (genus) to which the item belongs, and the differentiation (differentia) of the item from all other members of its class:

Item	Class	Differentiation
acceleration	the rate of change	of velocity with respect to time
resistance	any force	that tends to oppose or retard motion
glider	a light, engine-less aircraft	with lift surfaces and extended wings, designed for long periods of flight after launch from a towing vehicle

Item	Class	Differentiation
evaporation	a process	that changes a liquid into a vapor or gas
engineering	the application	of scientific principles to practical ends such as the design, construction, and operation of efficient and economical structures, equipment, and systems
ammeter	an instrument	that measures electric current's rate of flow
technique	a systematic procedure	used to accomplish a complex or scientific task
electricity	a physical phenomenon	arising from the existence and interactions of charged particles
carburetor	a mixing chamber	used in gasoline engines to produce an efficient explosive vapor of fuel and air

CLASSIFYING THE ITEM

As the preceding examples show, careful modification of the class goes a long way toward completing the definition. The narrower the class, the more meaning it conveys, and the less that needs to be said in the differentiation. Classifying an item as an "intricate device" accomplishes very little; the world is full of intricate devices, and the item still needs to be differentiated from all of them. Saying that an item "is what" or "occurs when" suggests that you are classifying when you obviously are not. Such words are sometimes useful in informal definitions, but they have no place in formal sentence definitions.

DIFFERENTIATING THE ITEM

If the differentiation applies to more than one member of the class, the definition lacks precision. Even a simple item like a ballpoint pen must be carefully differentiated from other members of its class; the ballpoint and other kinds of pens are distinguishable only by the design of their points. You must also guard against circular definitions, which repeat the very word being defined. The reader who needs help with *capacitance* will get nowhere if expected to understand the word *capacitor* in the differentiation. Noncrucial words such as *writing* in the term *technical writing* may, of course, be repeated. Occasionally, an extensive differentiation requires a second sentence.

Amplifying Formal Definitions

After reading a formal sentence definition, an uninformed reader often needs further explanation to understand the item completely. Eight methods for amplifying definitions follow.

DERIVATION

The derivation (origin) of some words, combinations of words, and acronyms helps clarify their meaning. *Ammeter,* for example, derives from the combination of *ampere* and *meter. Scuba* is an acronym for *self-contained underwater breathing apparatus.*

EXPLICATION

In this context, *explication* means defining difficult words contained in the formal sentence definition. The words *velocity* and *charged particles* in the preceding formal definitions would need to be defined for many readers. When explicating, you can often write an informal definition rather than another formal one.

EXEMPLIFICATION

(MAN on the street)

Exemplification is a method of amplifying that gives readers something concrete to help them understand a term. Attempt to select examples that are familiar to a wide range of readers. For example, you would miss a good opportunity if you did not amplify the definition of *weightlessness* by referring to the astronauts' experience in space. In the following paragraph, a formal sentence definition of *thermal transfer* is amplified by an extended example.

Thermal transfer is a transmission process by which cold

things absorb heat from warmer things. For example, when ice

cubes are placed in a glass of room—temperature Coke, the ice cubes absorb heat from the Coke. The more heat the ice cubes absorb, the colder the Coke becomes. The cubes continue absorbing heat until they have completely melted. At this point, the Coke and the water from the melted ice have the same temperature. If the mixture of Coke and water is allowed to sit, however, thermal transfer continues: the mixture absorbs heat from the surrounding air. When the temperatures of the air and the mixture are equal, heat transfer stops, having produced a glass of diluted, room-temperature Coke.

ANALOGY

An analogy points out a similarity between otherwise dissimilar things. If something is unknown to readers, they will be aided if you call attention to its similarity, however slight, to something they do know. The *Supervisor's Handbook* for U.S. Steel Supply draws on the readers' knowledge of golf to explain a term: "The Standard Cost System is a control for comparing what we actually spend (the strokes we took) with what we should have spent (par)." The following definition shows that a chemical equation is analogous to a recipe.

A chemical equation is a symbolic representation of a chemical reaction. The symbols are separated by an arrow; the symbols to the left are called reactants and the ones to the right are called products. Below is the chemical equation for creating sodium chloride, or plain table salt:

$$Na + Cl + Heat \rightarrow NaCl$$

Sodium (Na) reacts with chlorine (Cl) when heat is applied,
forming NaCl (sodium chloride).

Chemical equations are like kitchen recipes in that both
identify ingredients that react to form a finished product.
For example, the recipe for bread can be stated in the following way:

$$Flour + Water + Yeast + Heat \rightarrow Bread$$

Heat and yeast cause flour and water to react, creating
bread.

COMPARISON-CONTRAST

A comparison-contrast definition shows both the likenesses of and differences
between similar items such as a dune buggy and a jeep. Like other methods
of amplifying, the comparison-contrast definition capitalizes on something the
readers know in order to explain something they do not know. Comparing
and contrasting a semiconductor of electricity with a conductor of electricity
works only if the readers know what a conductor is. In the following example,
circuit breaker is defined through a comparison and contrast with a better-
known item, a fuse.

A circuit breaker is an automatic protective device that
prevents electrical circuits from being overloaded. Over-
loads occur when too much current is drawn from a source; unless the circuit is broken, the excess current will overheat
the wiring, damaging it and possibly causing a fire. Therefore, breakers are installed at places in a circuit where over-
loading may occur, and they go into action the moment the

circuit becomes overloaded. A circuit breaker is similar to a fuse because both break electrical circuits by opening metal contacts. In a fuse, the contact is a piece of metal (the weakest point in the circuit) that melts and breaks when current becomes excessive. Circuit breakers, however, are composed of a movable contact, a spring, and a trip coil. Excessive current energizes the trip coil, causing the spring to pull the movable contact and open (break) the circuit. Also, blown fuses must be replaced but circuit breakers can be either manually or automatically reset.

CAUSE-AND-EFFECT

Some things are so elusive that they must be defined in terms of their causes and effects. For example, nobody really knows what electricity is, but the interactions that cause it can be explained. Magnetism, on the other hand, must be defined through the force it produces, its effect. The causes and effects of lift are described as follows to amplify the formal sentence definition.

Lift is the aerodynamic force by which the surfaces of an aircraft oppose the pull of gravity. An aircraft's major lift surfaces are the wings. Each wing's curvature (Fig. 1) causes air flowing beneath it to arrive at the trailing edge before air flowing above it. Therefore, the lower-surface flow attempts to lap around the trailing edge, creating a vortex (whirlpool). The vortex's rotation increases the ve-locity of upper-surface flow until it moves from leading to trailing edge as fast as the lower—surface flow does. This increase in velocity over the wing causes a decrease in pressure on top of the wing. Because pressure on the wing's upper surface is less than that on its lower surface, lift occurs, permitting the aircraft to gain and sustain altitude.

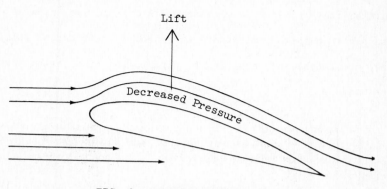

FIG. 1: AIRFLOW CREATING LIFT

ILLUSTRATION

Drawings and diagrams can be very helpful for reinforcing definitions. For example, a small, labeled drawing would help explain the function of a carburetor, and the definition of an abstract term such as resistance could be amplified by a diagram of forces in opposition. The preceding definition of lift is reinforced by an illustration of the airflow around an airfoil.

ANALYSIS

Analysis refers to the division of an item into its parts. This method aids readers' comprehension by allowing them to grasp the definition bit by bit. For example, dynamics becomes easier to understand when its two main parts, kinetics and kinematics, are discussed individually. The following paragraph analyzes and exemplifies two kinds of springs.

A spring is a taut but elastic body, generally constructed of brass, steel, or bronze, that yields under stress
and returns to its normal form when the stress is removed.
Two kinds of springs are spiral and helical. Spiral springs,
such as those used in clocks, are wound about a center and
furnish their power by unwinding. Helical springs, such as
those within retractable pens, are wound in cylindrical form;
they are capable of returning to normal size after being compressed.

The following pages contain three student papers that will serve as models.
Each begins with a formal sentence definition and uses several methods to
amplify the definition. Read the models carefully and identify the methods of
amplification.

Whether or not you are knowledgeable about the items being defined,
examine the models critically from your point of view. Are the formal sentence
definitions accurate? Would other methods of amplification help you understand? Which ones? Is information presented logically so that one idea prepares you to understand the next? Offer your class specific suggestions for
improving the models.

Models

SONAR

Sonar is an underwater detecting system that uses sound. The word <u>sonar</u> is derived from the words <u>so</u>und <u>n</u>avigation <u>and</u> <u>r</u>anging. Sonar is used for determining the depth of water and for locating and pinpointing underwater objects such as submarines.

derivation

In a sonar device, electrical signals are generated by a transmitter and sent to a transducer. The transducer con- verts the electrical signals into sound waves and sends them into the water. The waves travel until they hit an object, and then they bounce back to the transducer. At the same time sound waves are sent out, a signal is sent to a receiver that measures the time between the departure and the return of the signal. Because sound travels at a known speed in water, the distance of a particular object can be determined.

cause & effect

Sonar can be understood by examining a round trip in an automobile. If a car travels 60 mph, and a round trip to a certain city takes 1 hour, then the distance traveled is 60 miles. The distance to the city can easily be calculated to be 30 miles. With sonar, the round trip of sound waves per- mits the distance of an object to be measured.

Analogy

TURBOCHARGER

A turbocharger is a turbine—driven device that boosts the horsepower of an internal—combustion engine. The turbocharger forces air into an engine's cylinders at a pressure greater than atmospheric pressure. In effect, this increase in compression increases an engine's size. If a turbocharger forces 50% more air into an engine than it normally consumes, the engine acts as if its horsepower were 50% greater.

Cause & Effect

The name turbocharger derives from the words turbine and charge. In an internal—combustion engine, a turbine converts the energy of moving exhaust gases into mechanical power. To accomplish this, most turbochargers have two bladed wheels on the same shaft. The engine's exhaust gases travel past one of the wheels, spinning it and the shaft. Because the other wheel is on the same shaft, it also spins, acting like an air compressor and forcing air into the engine. Charger derives from the turbocharger's ability to charge an engine with air just as a bellows charges a fire with air.

derivation

explication

Turbochargers are installed on almost all USAC (United States Auto Club) Indianapolis racers. Street automobiles such as the 1978 Buick Regal V—6 and some models of Saab and Porsche also contain turbocharged engines. Other manufacturers might also develop turbochargers as the demand for efficient, small—engine automobiles grows.

FAIL-SAFE

Fail-safe is the condition of a system able to correct its own malfunction or shut itself off before any damage can result. The term <u>fail-safe</u> is used because if part of the system fails, causing an unsafe operating condition, the system automatically takes steps to return to a safe condition.

derivation

Fuses and circuit breakers are common fail-safe devices used to prevent electrical wiring from overheating. When a fuse blows, it stops the flow of electricity; circuit breakers do the same thing, but some of them are also capable of automatically restarting the flow of electricity once the danger of overheating has passed. These and other fail-safe systems, such as elevators, achieve safety by simply stopping.

exemplification

Systems such as aircraft, however, cannot achieve safety by merely stopping. In order to be fail-safe, they must trigger back-up devices in case the main system malfunctions. By automatically switching to the back-up devices, these systems bypass the malfunctioning parts and continue to operate safely.

Writing Assignment

Write definitions of four items for an uninformed reader, beginning each of them with a formal sentence definition.

1. Write a 100-word definition (one paragraph) using analogy as the major method of amplification.
2. Write a 100-word definition (one paragraph) using comparison-contrast as the major method.
3. Write a 100-word definition (one paragraph) using cause-and-effect as the major method.
4. Write a 200-word definition (two or more paragraphs) using four of the eight methods of amplification discussed in the chapter. Immediately after you have used a method, identify it in brackets. Use only methods that will do your reader some good. For example, every term has a derivation, but that information is not always valuable to a reader.

In writing the 200-word definition, you will find that using one method leads you to another. Consider providing synonyms and other informal definitions in each of the four definitions, but do not count them among the four methods in your 200-word definition.

You may not have to write many definitions of this length in industry, but mastery of the technique will prove useful in writing all reports. As was emphasized early in this chapter, you are in the process of becoming a specialist in your field; the need to define specialized terms goes with the territory.

Try to select terms, probably from your technical area, that do not require you to look up the definition. If you use sources, however, give credit to them. Some possible topics are the following:

hypothesis	Styrofoam
compression	objectivity
microwave oven	horsepower
galvanizing	calibration
equation	feedback

Exercises

1. Working with other members of the class who are majoring in your technical area, select two terms that are commonly used and understood in your laboratories. As a group, write formal sentence definitions of the terms. Then ask the other groups to provide oral definitions of the terms. The purpose of this experiment is to determine how knowledgeable people in other technologies are about terminology used in your technical area.

2. Write formal sentence definitions of the following six terms, assuming that your reader has no familiarity with them. As you know, first you must place the item into its general class, and then you must differentiate it from all other members of its class.

profit	kite
algebra	hockey puck
quarterback	friction

As you classify each item, avoid saying "is what" because those are not words of classification. Also, consider using modifiers to narrow the class, thus reducing the number of words necessary in the differentia. Write each formal sentence definition very carefully, remembering that its function is to distinguish an item from every other item in the world.

3 Describing a Mechanism

DESCRIPTION is more widely used than any other technical writing technique. In fact, you will have to do some describing in every report you write. A mechanism's physical characteristics must be understood before its operation can be grasped. What size is it? What material? How is it powered? Can it take a beating? These and 100 other questions require answers.

Description is most frequently used in manuals, where the writer provides a detailed physical description of the mechanism before explaining how it is operated. In addition to manuals, however, descriptive sections are necessary in proposals, feasibility reports, and short, informal reports. The feasibility writer, who has been asked to report on several alternative systems, describes and compares them before recommending one for installation. In proposals, the writer tries to sell the firm's design by providing the potential buyer with a thorough description of it. These reports cannot contain vague or ambiguous writing. The descriptions must be detailed, clear, and accurate to make readers feel secure in their knowledge of the mechanism.

The All-inclusive Mechanism

A *mechanism* is any system of parts that operates in a definable way. In other words, not only a machine but almost anything that has a specific function can properly be called a mechanism. The following description of an airplane hangar demonstrates the inclusiveness of the word *mechanism*. The quotation is from a *Design News* article (February 3, 1969) entitled "Hyperbolic Paraboloid of Steel Forms" by Lars Soderholm. Because the readers of *Design News* are professionals in the construction field, Soderholm does not have to define such terms as *fascias, struts,* and *purlins.* These readers, however, are *not* familiar with the newly designed hangar, so Soderholm must describe the hangar clearly enough that they can visualize it:

> The hangar is supported by front and rear fascias built in the form of canted A-frames with their legs resting on huge concrete buttresses. Longitudinal tension bars are draped between the fascias and serve to support the roof almost in the same manner as cables from a suspension bridge. Wide-flange compression struts or purlins run diagonally between the fascias to prevent them from bending inward. They also provide vertical support for the roof decking and snow loads.

The front fascia, constructed of wide-flange beams, holds the nose section of the aircraft. The rear fascia is made up of a truss section and will cover two king-sized, self-propelled doors that swing open on a steel track, offering a 164-ft. opening. (Pp. 26–27)

The article is evidence that each technical discipline in our highly technical society contains items that require description. Readers in all fields need to know how things are constructed in order to clearly understand how they work. For this reason, the word *mechanism* will be used comprehensively throughout this chapter and Chapter 4. The principles of description can usefully be applied to items in such diverse fields as chemistry, physics, biology, electrical technology, mechanical technology, social science, speech and hearing correction, nursing, and geography. The compass, insulin syringe, and lightbulb described at the end of this chapter exemplify the various items that lend themselves to physical description.

Reaching the Reader

A mechanism must be described as carefully as possible to achieve communication with an uninformed reader. First, you have to become thoroughly familiar with the mechanism yourself. If possible, get your hands on it, take it apart, and examine it closely. You have won most of the battle if you know the mechanism inside out. On the other hand, you can become a victim of your own knowledge. It is easy to overlook things that seem obvious to you but are not at all obvious to your reader. If you accidentally omit an essential fact, the reader does not have a chance.

Communication also requires definitions of terms unfamiliar to your reader. The *Design News* article quoted earlier is aimed at people with experience in the construction field, but the author uses methods of defining explained in Chapter 2. When he says that the hangar's tension bars "support the roof almost in the same manner as cables from a suspension bridge," he is obviously comparing something unknown to his readers (the hangar design) to something known to them (a suspension bridge). In addition to this, the author makes an analogy when he uses the word *A-frames* in his opening sentence; he points out a similarity between the shape of the frames and the shape of something known to all readers, the first letter of the alphabet.

Another way of reaching the uninformed reader is through the use of visuals. The article in *Design News* contains diagrams and photographs backing up the written description. Such visuals reinforce the description of a complex mechanism, and the key word is *reinforce*. Visuals aid or supplement the description; they should never stand alone. Also, using a visual simply because it looks good is as defeating as using words that only sound good. Visuals should be reserved for occasions when they will assist the effort to communicate. The type of visual you select depends on the mechanism and the reader. A cross-sectional drawing may be better than an exploded view for clarifying a particular piece of equipment. Your options concerning visual material are discussed in detail in Chapter 7.

Outline for Description of a Mechanism

Guidelines for describing a mechanism are presented in the following outline. For demonstration purposes, the description section of the outline assumes that main part B has two subparts.

I. Introduction
   A. Definition and Purpose
   B. Overall Description (size, weight, shape, material)
   C. Main Parts
II. Description
   A. Main Part A (definition followed by detailed description of size, shape, material, location, and method of attachment)
   B. Main Part B (definition followed by overall description, and then identification of subparts)
      1. Subpart X (definition followed by detailed description of size, shape, material, location, and method of attachment)
      2. Subpart Y (same as for subpart X)
   C, D, etc.: Other Main Parts (same as for main part A or main part B)

I. INTRODUCTION

An introduction provides an overview of the mechanism, and its importance cannot be overemphasized. If a reader becomes lost at the point of departure, he or she will never get through the remainder of the description. For that reason, the introduction contains three sections designed to send the reader into the body of the paper with enough general information to grasp the specifics of the mechanism.

A. Definition and Purpose   Because the reader is uninformed, the first sentence in the paper must be a formal sentence definition of the kind described in Chapter 2. If that sentence requires amplification, use one or more of the eight methods contained in Chapter 2. After amplifying, provide concrete information about the mechanism's purpose, giving the reader a fuller understanding of where, when, and why the mechanism is used. The sample definition section that follows is from a description of a soldering iron. The section contains only one paragraph, and the formal sentence definition serves as a topic sentence.

INTRODUCTION

Definition and Purpose

The soldering iron is an electrical hand tool used to form a junction between electrical wires or between wires and components. The iron is used to heat these metal pieces so that solder (a tin and lead substance) will melt around them. When the solder cools and hardens, it forms a permanent electrical connection. The soldering iron's main application is in the assembly of electrical or electronic equipment such as televisions and stereo receivers.

B. Overall Description   Obviously, a reader must be informed about a mechanism's overall appearance before he or she can understand its individual parts. For your paragraph of overall description, select appropriate information from the following physical characteristics:

Size and weight: If various sizes are manufactured, inform the reader of the range available, and then give dimensions for the particular size you have chosen to describe. Also provide the weight of the mechanism, if pertinent.
Shape: You can frequently clarify the mechanism's overall shape by using an

analogy such as *A-shaped* or *pencil-shaped*. Consider reinforcing this information by presenting a visual in either the overall-description or the main-parts section of the introduction.

Material: If the entire mechanism is made of only one material such as polished steel, tell the reader in this section; however, if its parts are of different materials, save this specific information until the body of the paper.

The introduction is your last opportunity to provide facts about the mechanism as a whole. Therefore, do you hesitate to give the reader additional general information that you believe will be helpful. The following sample overall description presents several such facts.

```
Overall Description

     The soldering iron described in this paper is a 25-watt

Weller Marksman, Model SP-23.  The iron has a length of 8½

inches, a maximum diameter of 1½ inches, and a weight of

1-3/4 ounces.  A cord extends from its handle for plugging

into a standard 110-volt wall outlet.  The iron develops a

tip temperature of 750 degrees Fahrenheit.
```

C. *Main Parts*   The main-parts section is much more important than its brevity would suggest. In this section, you simply identify the mechanism's main parts, but you must identify them *in the order they will be described in the body of the paper*. What order will you use? Exterior to interior? Interior to exterior? Most important to least important? Left to right? Top to bottom? Some combination of these? The answer depends on the mechanism, and mechanisms fortunately lend themselves to logical analysis. You need to select an order that will aid instead of confuse your reader in the body of the paper. If the part you describe first forces you to make several references to other parts, your reader will become lost because she or he as yet knows very little about those other parts.

In other words, the main-parts section controls your organization of the rest of the paper, just as a scope section controls the organization of an investigative report. In any writing, you want to break your information into major sections (main parts) so the reader can assimilate the material bit by bit. Then you want to present those major sections in a logical order so your reader can easily follow the information as you present it. Do not hesitate to reorganize your physical description if you discover that the order you originally selected is not effective.

Main Parts

The three main parts of the soldering iron are the handle, heating element, and tip (Fig. 1).

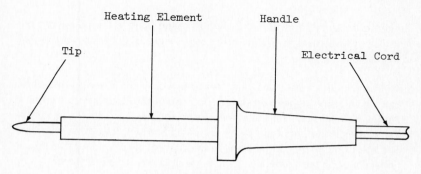

FIG. 1: SOLDERING IRON

II. DESCRIPTION

In the body of the paper, the detailed description, a main part receives a heading and a single paragraph. Provide either a formal sentence definition or an informal, operational definition as the topic sentence for each paragraph. Then complete the paragraph by describing the main part's physical characteristics.

A. *Main Part A*   Let's examine such a paragraph from the body of the soldering iron paper just begun.

DESCRIPTION

Handle

The handle serves as an insulator against the intense heat of the tip and as a gripping surface for the operator.

```
It is a 5-inch-long plastic cone with a 1½-inch-diameter col-

lar at its wide end and an opening for the electrical cord at

the other end.  The inside of the cone is hollow, allowing

the electrical wires to pass through it and into the heating

element.  Three screws connect the entire handle to a disk on

the heating element.
```

After an informal, operational definition of the handle, the paragraph presents its physical characteristics of size, shape, material, location, and method of attachment:

Size: The handle's length and diameter give the reader a perspective of the part. In industry, depending on the purpose of the description, extremely thorough dimensions are sometimes necessary; they are placed on a drawing of the part to avoid cluttering the paragraph with too many numbers. The use of numbers and hyphens in the preceding paragraph conforms to generally accepted guidelines, which are discussed in the appendix.

Shape: The word *cone* clarifies the shape of the handle, just as the words *cylinder, disk,* and *rectangle* would indicate other shapes. Simple analogies comparing shapes to letters of the alphabet (A, C, L, T, U) are common in technical writing. Visuals of particularly complex parts are also frequently presented to reinforce the written description. When this is done, the visual must be numbered, titled, and referred to in the paragraph, as was done in the main-parts section of the introduction shown earlier.

Material: Early in the paragraph plastic is identified as the material of the handle.

Location and Method of Attachment: *Location* refers to the position of the part in relation to the other parts: the handle is located between the soldering iron's cord and its heating element. The paragraph's final sentence states that screws attach the handle to a disk on the heating element.

Clear description requires not only identifying the physical characteristics but stating them in a readable, coherent paragraph. You may need several revisions before deciding on the most logical order for presenting your information.

B. Main Part B When describing the parts of a mechanism, you may find that you cannot explain a complex part clearly in one paragraph. This means that you should divide the part into subparts. Doing so will simplify your writing task and improve your reader's understanding. As was shown earlier in the outline, you should provide a brief paragraph to define the main part, describe it generally, and introduce the subparts. Then you should give each of the subparts a heading, define it, and describe it in detail. In the sample that follows, the soldering iron's heating element is divided into subparts that are described individually. Then, to complete the sample description of a soldering iron, the final main part of the mechanism is described.

Heating Element

    The heating element changes electrical energy, transmitted from the wall outlet, into heat. It is 2½ inches long and 3/8 inch in diameter, and it is made up of a casing and a coil.

    Casing

        The casing is a stainless steel cylinder that encloses the heating coil. One end of the cylinder is welded to a metal disk, and screws connect the disk to the handle. At the other end of the cylinder is a copper-alloy threaded socket that accommodates the soldering iron's tip.

    Coil

        The heating coil receives electricity transmitted by the wires inside the handle. The coil is made up of fine electrical wire wound around a steel core. When

the soldering iron is plugged in, the coil becomes hot,

just as the wires in a toaster do.

C. Main Part C   The soldering iron's third main part, the tip, does not require division into subparts. Therefore, the tip is described in the same manner as the first main part, the handle.

Tip

The stainless steel tip, which is 1–3/4 inches long,

transfers heat from the heating coil to the junction to be

soldered. One end of the tip is threaded so it can be

screwed into the casing's socket to make contact with the

heating core. The other end is beveled into two flat sur-

faces similar to those on the end of a screwdriver. These

surfaces are nickel plated to withstand extremely high tem-

peratures.

A mechanism is the sum of its parts, and detailed descriptions of each add up to a detailed description of the entire mechanism. Physical descriptions generally require no conclusion because a good description leaves nothing to conclude.

Description is the most widely used technique in technical writing. In industry you will never write a report that does not require some description, and you may occasionally write one that contains nothing but description.

As you read the three student models that follow, observe their organization. Your analysis of a mechanism and the order in which you describe its parts will greatly affect your ability to communicate. Also, offer your class specific suggestions for improving the models. Are all the parts adequately described in the description of a bow compass? In what order would you have described the parts of the incandescent lamp? What are the strengths and weaknesses of the papers? Be critical of them, and become just as critical of your own writing.

Models

DESCRIPTION OF A BOW COMPASS

INTRODUCTION

Definition and Purpose

A bow compass is a drafting instrument used to draw circles or arcs. It is also used to mark off distances prior to
the drawing of lines. Mechanical artists use the compass to
create architectural and technical drawings that may be re-
produced as blueprints.

Overall Description

Bow compasses are A-shaped and manufactured in a wide
range of sizes. The compass described in this paper is 3½
inches tall when held in its drawing position, perpendicular
to and with both points on the drawing surface. The compass
is made of polished alloy metals and weighs about 3 ounces.

Main Parts

The main parts of the bow compass are the handle-and-re-
tainer assembly, pivot washer, adjusting assembly, and legs
(Fig. 1).

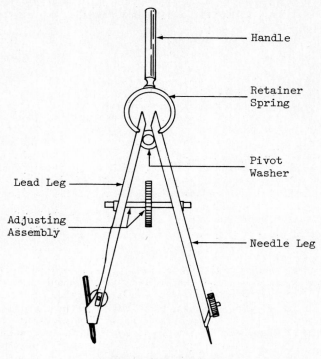

FIG. 1: BOW COMPASS

DESCRIPTION

Handle—and—Retainer Assembly

    The handle—and—retainer assembly, located at the top of
the compass, consists of a handle connected to a ring—shaped
retainer spring. The handle is 1/8 inch in diameter and
13/16 inch long. Along its length, ribs are scribed to permit sure handling by the mechanical artist. The handle
screws into a 1/32—inch threaded hole in the retainer spring.
This spring is 9/16 inch in diameter with a 3/16—inch opening

at its bottom. When the legs of the compass are inserted into this opening, the spring holds them in place (retains them).

Pivot Washer

The pivot washer, located immediately below the retainer spring, is held in place by the legs. As the artist narrows or widens the legs, they pivot on this finely machined, $\frac{1}{4}$-inch-diameter washer.

Adjusting Assembly

The adjusting assembly consists of a 3/8-inch-diameter disk at the center of a threaded axle that is 1/16 inch diameter by 7/8 inch long. One end of the axle has right-handed threads and the other has left-handed threads. The ends extend through the legs and are held there by two nuts. Enclosed by the legs, these nuts do not move; however, when the artist turns the adjusting disk, the axle rotates, moves through the nuts, and adjusts the width of the legs.

Legs

The legs have very similar physical characteristics. Both are $2\frac{1}{4}$ inches long and have small, circular openings into which the ends of the retainer spring fit. On the inside of each leg, beginning about 1/8 inch from the top, is a 3/16-inch vertical slot into which the pivot washer fits.

Five-eighths of an inch from the top of each leg is a threaded hole occupied by the axle of the adjusting assembly. One leg holds the needle and the other holds the lead.

Needle Leg

On the inside of the needle leg, directly below the adjusting axle, is a 1/16-inch-wide slot extending down to the end of the leg. The needle, when placed in this slot, slides through a 1/32-inch hole in a pin. Then, on the outside of the leg, a disk-shaped nut screws onto the pin and holds the needle in place.

Lead Leg

The outside of the lead leg has a 1/32-inch-wide slot that helps hold the lead. Near the bottom of the leg, a 1/16-inch hole is drilled through the slotted halves for the tightening bolt. When a nut is tightened onto this bolt, it pulls the slotted halves together, securing the lead.

DESCRIPTION OF AN INSULIN SYRINGE

INTRODUCTION

Definition and Purpose

An insulin syringe is a medical instrument used for injecting insulin into the body. Physicians, nurses, or dia-
betic patients themselves use the syringe to add insulin that
is lacking in the bodies of diabetics.

Overall Description

Insulin syringes are manufactured in one standard size.
The syringe is $6\frac{1}{2}$ inches tall when held with the plunger
fully withdrawn (open) and $4\frac{1}{4}$ inches tall when the plunger is
enclosed (closed) in the barrel.

Main Parts

The syringe, as Fig. 1 shows, is composed of a plunger
assembly, barrel, and needle.

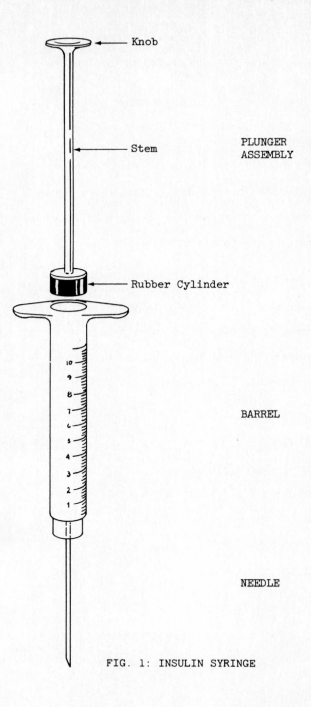

Knob

Stem

PLUNGER
ASSEMBLY

Rubber Cylinder

BARREL

NEEDLE

FIG. 1: INSULIN SYRINGE

DESCRIPTION

Plunger Assembly

The plunger assembly is used to push the insulin through the barrel and into the needle. It is located at the top of the syringe and consists of a knob, stem, and rubber cylinder, all of which are molded together.

Knob

Located at the top of the plunger assembly, the plastic knob is gripped to move the plunger. The knob is a smooth disk having a diameter of 3/8 inch and a thickness of 1/32 inch.

Stem

The stem allows the rubber cylinder, attached to the bottom of the stem, to extend to the bottom of the barrel when the plunger is fully closed. This forces the insulin out of the barrel and into the needle. The stem is made of plastic and has ribs scribed along its length. It measures $3\frac{1}{4}$ inches long and 1/8 inch in diameter.

Rubber Cylinder

The rubber cylinder attached to the lower end of the stem fits snugly into the barrel, allowing no air or foreign particles to enter the insulin. The cylinder is ¼-inch tall and has a 3/16-inch diameter.

Barrel

The barrel of the syringe holds the insulin prior to its injection into the body. A handle is molded to the top of the barrel, providing better gripping of the syringe during injection. The handle is ½-inch long, 3/8-inch wide, and 1/8-inch thick. The smooth, clear plastic barrel is marked off in cubic centimeter intervals. This 3-3/8-inch tall barrel has a diameter of 3/16 inch. A 1/64-inch opening at the bottom of the barrel allows insertion of the needle.

Needle

The needle is the metal part of the syringe that penetrates the patient's skin. The insulin travels through the needle and into the body. The needle is held in place by a plastic hub that is molded to the underside of the barrel. The needle is a hollow cylinder 5/8-inch long and 1/64-inch in diameter. The end of the cylinder has a slanted opening and an extremely sharp point (Fig. 2).

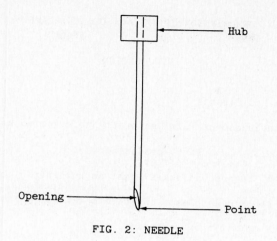

FIG. 2: NEEDLE

DESCRIPTION OF AN INCANDESCENT LAMP

INTRODUCTION

Definition and Purpose

The incandescent lamp, generally referred to as an electric lightbulb, is a light-producing device. It is called incandescent because it emits light as a result of being heated; a fluorescent lamp, on the other hand, creates light through the emission of radiation.

Overall Description

Shapes and sizes of lightbulbs vary, but most are pear shaped, approximately 4 inches tall and $2\frac{1}{2}$ inches in diameter. The electricity necessary to power a lightbulb is its wattage. Standard wattages are 25, 40, 60, 75, 100, 150 and 200, although other wattages are available. Most lightbulbs have a life of approximately 750 hours.

Main Parts

The main parts of a lightbulb are the base, lead-in wires, interior support, filament, and bulb (Fig. 1).

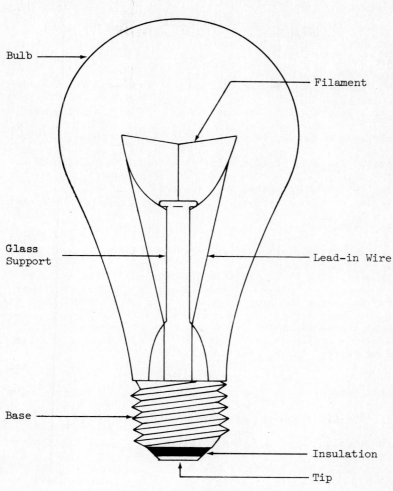

Bulb

Filament

Glass
Support

Lead-in Wire

Base

Insulation

Tip

FIG. 1: INCANDESCENT LAMP

DESCRIPTION

Base

The base is the threaded part of the lamp, which screws into a lamp socket. The base is made of a conductive metal (aluminum or brass) to which one of the lead-in wires is soldered. The other lead-in wire is soldered to the tip of the base, which is insulated from the rest.

Lead-In Wires

The lead-in wires in the lightbulb carry current to the filament. The wires are made of a heat-resistant metal so they can carry the lamp's required current. Connected to the base, the wires extend up to the filament, passing through a glass support in the center of the bulb.

Glass Support

The glass support is a tube that holds the lead-in wires and filament in place. Resembling a test tube, the support is molded to the bottom of the bulb and extends up to the center of the bulb. Three tiny wires are connected to the support to keep the filament from sagging.

Filament

The filament is a flexible thread of tungsten that heats and glows as electricity passes through it. The filament reaches a white heat because of tungsten's resistance to the flow of electricity. Tungsten also has a high melting point (above 6000 degrees Fahrenheit), making it an excellent material for the filament. The filament is held in the center of the bulb by the interior support, and its ends are soldered to the ends of the lead-in wires.

Bulb

The bulb is a glass enclosure that protects the filament. Exposure to air would cause rapid oxidation (burning away) of the filament, so a neutral gas (argon or nitrogen) is sealed into the bulb. Molded into the base, the bulb can be either clear glass to produce brilliant white light, or it can be coated with a phosphorous substance to create a soft white light. Molecules of hot tungsten slowly darken the bulb's interior surface as the filament deteriorates.

Writing Assignment

Write a 500-word physical description of a mechanism for an uninformed reader. Select a small mechanism that you have worked with in your technical area, in the science laboratory, on the job, or around the house (workbench, kitchen, garage). Some possible topics are listed, but you are the best judge of topics; review your own experience and select a topic about which you are knowledgeable.

carburetor	thermostat
mechanical pencil	diode
fuel pump	core module from a computer
gyroscope	tachometer
gate valve	opaque projector

Within your paper, you will inevitably refer to the mechanism's operation, particularly when you are defining parts. Make sure, however, that your paper emphasizes the mechanism's physical characteristics rather than its operation. Describing a mechanism's operation is a separate technique of technical writing, which will be explained in Chapter 4.

Exercises

1. The frame of a dragster, like the frame of any car, is composed of side rails and crosspieces. However, the dragster differs from conventional cars in three main ways: (1) It is made of very light, somewhat springy material to transfer weight to the back wheels at the beginning of the ¼-mile journey. (2) Its rear axle, rather than being suspended from the frame, is welded to the frame. (3) The driver's compartment holds only one person and has a roll cage to protect him or her. The frame, shown in the figure, serves as a skeleton that holds every part of the completed dragster.

Write a detailed description of the dragster's frame. The rear axle housing has been excluded from the drawing to make the frame easier to describe. Aim your description at a reader who has never seen a dragster. Rely solely on words to achieve communication: provide no visuals and assume that your reader does not have access to the drawing.

The following is some additional information about the dragster's frame:

length: 17 feet
width: front crosspiece—30 inches, rear crosspiece—26 inches
weight: 100 pounds
material: seamless steel (aircraft) tubing
method of construction: welding

2. A cigarette lighter's three main parts are its outer casing, inner casing, and spring-and-screw assembly:

Outer Casing	Inner Casing	Spring-and-Screw Assembly
a. case	a. case	a. spring
b. cover	b. wind baffle	b. screw
c. hinge	c. striking wheel	c. plug
	d. latching device	

Write a detailed description of the outer casing and inner casing for an uninformed reader. Clarify the location and shape of the lighter's metal parts, but do not describe the wick, flint, or cotton filler. If possible, examine a lighter closely before beginning your description.

The following description of the spring-and-screw assembly will give you an idea of the amount of detail your description should contain:

The spring-and-screw assembly holds the flint against the striking wheel. The assembly consists of a spring, a screw, and a cylindrical plug. The spring is $1\frac{1}{4}$ inches long and 1/8 inch in diameter. Attached to one end of the spring is a plug, which is 3/16 inch long and 1/8 inch in diameter. The other end of the spring connects to a $\frac{1}{4}$-inch-long metal screw that has a slot across its top. The entire assembly fits plug first into a threaded tube within the inner casing of the lighter. When the screw is turned into the end of the tube, it causes tension in the spring, pushing the flint securely against the striking wheel.

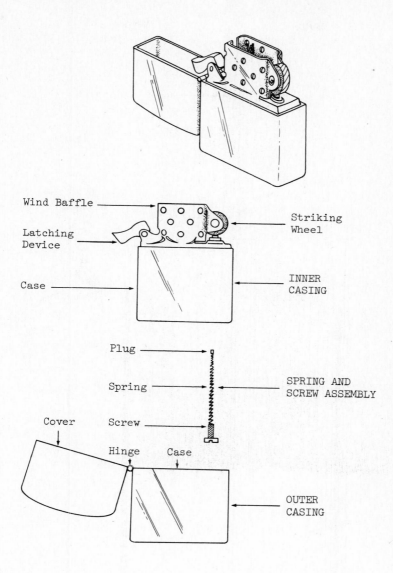

3. Write a detailed description of a ballpoint pen for an uninformed reader, emphasizing the shape and dimensions of each of the parts.

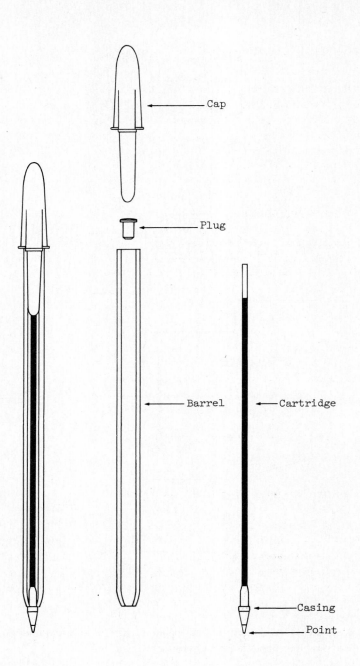

4 Describing a Mechanism in Operation

IN today's technical society, people depend on the action of machines. Processes once accomplished by workers with tools are now completed by machines under the control of workers, and today's technical writing reflects that shift. Describing a mechanism in operation has become a more common writing technique than the conventional process description that emphasizes the role that people play in technology. To put it another way, more and more buttons are being pushed all the time, and the action that occurs after the button is pushed takes precedence over the human activity of pushing the button.

Therefore, this chapter emphasizes the techniques for describing mechanisms in operation. The more traditional process description will, however, be discussed and exemplified toward the end of the chapter.

Again, the word *mechanism* includes all systems whose parts operate in a definable way. Thus, the method used for describing a mechanism's operation is applicable to almost all disciplines. Even nonhardware items, such as athletic teams, can be viewed as mechanisms. Successful teams are commonly referred to as well-oiled machines or big red machines, and the comparison is valid. The power sweep, a play used by most football teams, involves 11 integral parts (players), each of whom functions with precision when the play is successful. Each player has an important job, if only to act as a decoy. The function of a power sweep is to gain yardage, and ultimately to score a touchdown. Also, most teams subscribe to Vince Lombardi's run-to-daylight principle of operation, which means that the ball carrier heads upfield at the first opening. Although this example is unusual, it should serve to emphasize the wide applicability of operational descriptions.

Outline for Describing a Mechanism in Operation

A description of a mechanism in operation should not be confused with a physical description of the mechanism. The two require separate techniques of report writing, and (as was mentioned in Chapter 3) the physical description of a mechanism generally precedes a description of its operation in industrial writing. The following outline is suitable for describing virtually any operation.

I. Introduction
   A. Definition of Operation (also amplification of definition and general information about operation)

I. INTRODUCTION

The introduction to a description of operation contains general information designed to prepare the reader for the details that will follow. This requires a definition of the mechanism's operation, an explanation of its principles of operation, and an identification of the major sequences in the operation.

A. Definition of Operation   Logic tells us that an operational definition is appropriate in defining an operation. As was shown in Chapter 2, operational definitions differ from formal definitions in that no attempt is made to classify the item being defined; the emphasis is on what the item *does* rather than what it *is*. For example, an operational definition of a carburetor would read, "A carburetor delivers the proper ratio of gasoline and air to the cylinders of an internal-combustion engine." An operational definition of a gate valve would state, "A gate valve controls the flow of liquid through a pipe."

By itself, an informal, operational definition does not give the reader a firm enough basis on which to understand the specifics presented later in the paper. The writer must amplify, clarifying unclear terms and ideas contained in the definition, and then provide any general information that will assist the reader. The first sentence of the following sample definition-of-operation paragraph is an accurate operational definition. Nevertheless, two sentences are used to explain what "energy" and "power a vehicle" really mean, and a final sentence provides additional introductory information about pistons.

INTRODUCTION

Definition of Operation

The pistons of an internal—combustion engine produce the

energy necessary to power a vehicle. Reciprocal (up—and—

down) movement of the pistons causes rotation of the crankshaft. This mechanical energy, transmitted through a power
train (includes the transmission, drive shaft, axle), turns
a vehicle's wheels. Most automobiles have four, six, or
eight pistons powering the crankshaft.

B. Principles of Operation   All mechanisms have at least one major principle of operation that helps to clarify the "why" of the mechanism. Your readers will find it easier to understand the description of what happens in the body of your paper if they gain an idea of why it happens in your introduction.

For example, vaporization plays an important role in the operation of a carburetor, so a one-paragraph explanation of the principle would be appropriate in the introduction: "A carburetor operates according to the principle of vaporization. The carburetor distributes gasoline in the form of tiny droplets in an airstream. As a result of heat absorption on the way to the cylinders, these droplets vaporize, making the mixture into a flammable gas."

High-powered theoretical knowledge is not necessary to write such a paragraph. In fact, you should avoid complex, textbook-type presentations of laws and equations. All your reader needs to know is how the principle applies to the operation at hand. In effect, you are writing a limited, informal definition of the principle, so use the methods of defining discussed in Chapter 2. Here is the principle-of-operation section from the sample description of the operation of a piston used throughout this chapter.

Principle of Operation

Pistons operate according to the principle of energy
conversion. Combustion (burning of gases) within the piston
cylinders creates thermal energy (heat). Movement of the
pistons converts this thermal energy into mechanical energy
(turning of the crankshaft) to power the vehicle.

Sequences of Operation   Any operation can be broken into sequences. In a description of operation, the mechanism is analyzed in terms of sequences of operation just as in a physical description it is analyzed in terms of its main parts. The following list shows the strong relationship between the parts of a refrigerator and its operation, but because the column on the right indicates action, only its terms are appropriate for an operational description:

Physical Parts	Operational Sequences
compressor	compression
condenser	condensation
evaporator	evaporation

Many mechanisms have obvious sequences of operation, but others require careful examination in order to determine the most logical division. To make such a determination, view the mechanism's entire operation as a total chain-reaction sequence and decide where that total sequence can most logically be divided. Then, in the sequences-of-operation section, identify the sequences in chronological order.

The sequences you designate will become headings in the body of the paper, so this section of the introduction serves as a statement of scope. Most reports in business and industry contain a section on scope to tell the reader in advance what topics are covered in the remainder of the report. A sample sequences-of-operation section for a description of operation follows:

Sequences of Operation

In a four-cycle engine, four strokes of the piston are necessary to complete a cycle. The sequences of operation consist of the intake, compression, power, and exhaust strokes (Fig. 1).

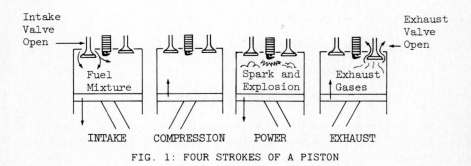

Intake Valve Open →   Exhaust Valve → Open

Fuel Mixture   Spark and Explosion   Exhaust Gases

INTAKE   COMPRESSION   POWER   EXHAUST

FIG. 1: FOUR STROKES OF A PISTON

II. DESCRIPTION OF OPERATION

In the description, or body of the paper, write one paragraph for each sequence of operation. The paragraph should begin with an operational definition of the sequence. This topic sentence states the purpose of the sequence, and the remainder of the paragraph then explains in detail the action necessary to achieve that purpose. The following passage is a description of a piston's intake-stroke sequence.

```
              DESCRIPTION OF OPERATION

Intake Stroke

        The intake stroke draws a fuel—air mixture from the car—
buretor into the piston cylinder.  During this stroke, the
piston moves from the top of the cylinder to the bottom.
This downward action creates a partial vacuum above the pis—
ton.  The vacuum draws the fuel—air mixture through the in—
take port and into the cylinder.  As the piston reaches the
bottom, the intake port closes, locking the fuel—air mixture
in the cylinder.
```

This detailed description clarifies the chain reaction of the mechanism's operation: the result of one action becomes the cause of another. Such description amounts to a series of cause-and-effect statements, as follows:

Cause	*Effect*
The intake stroke	draws a fuel-air mixture from the carburetor into the piston cylinder.
During this stroke, the piston moves from the top of the cylinder to the bottom.	This downward action creates a partial vacuum above the piston.

Cause	Effect
The vacuum	draws the fuel-air mixture through the intake port and into the cylinder.
As the piston reaches the bottom, the intake port closes,	locking the fuel-air mixture in the cylinder.

The major occurrence during the intake-stroke sequence is the creation of a vacuum. The downward action of the piston (cause) creates the vacuum (effect). In addition to being an effect, the vacuum is itself a cause: the vacuum (cause) draws the fuel-air mixture into the cylinder (effect). Clear explanation of cause-and-effect relationships is one of the qualities of good operational description.

Active voice also contributes to the clarity of an operational description. In fact, logic demands that you use active voice when you describe an action. *Voice* refers to the relationship between the subject and verb of a sentence. This relationship, which involves the use of transitive and intransitive verbs, is covered in the appendix. For now, however, let's analyze the intake-stroke paragraph according to the following abbreviated explanation of voice: When a sentence is in active voice, the subject *does* something. If the sentence is in passive voice, the subject *does nothing*; instead of acting, the subject is often acted on.

Active
The intake stroke (subject) draws (verb) a fuel-air mixture into the cylinder.

This sentence is in active voice because the subject (the intake stroke) acts; it draws the mixture into the cylinder. This information can also be stated in passive voice, but doing so increases the number of words and reduces the clarity.

Passive
A fuel-air mixture is drawn into the cylinder by the intake stroke.

Here the subject (the mixture) is passive; it is acted on by the intake stroke. Notice that the potential for active voice exists in this passive sentence: the words by the intake stroke indicate the real actor, the intake stroke. If intake stroke is made the subject of the sentence instead of the object of the preposition by, the sentence will become active.

Active
During this stroke, the piston (subject) moves (verb) from the top of the cylinder . . . .

Active
This downward action (subject) creates (verb) a partial vacuum . . . .

Active
The vacuum (subject) draws (verb) the fuel-air mixture . . . .

Active
As the piston reaches the bottom, the intake port (subject) closes (verb), locking the fuel-air mixture . . . .

Active voice pinpoints the action of a mechanism's parts, making their operation easier to grasp. Passive voice, on the other hand, wastes words and often confuses the meaning of a sentence. Whenever possible, use the active voice for descriptions of operations and for most other kinds of technical writing.

The following three sequence-of-operation paragraphs demonstrate the use of cause-and-effect statements and active voice. Notice also that a topic sentence, an operational definition of the sequence, states the central idea of each paragraph.

Compression Stroke

The compression stroke traps the fuel—air mixture against the top of the cylinder. As the piston moves up, it forces the mixture to occupy less space. This compression of the fuel—air molecules raises their temperature, making them volatile.

Power Stroke

    During the power stroke, combustion of the compressed fuel—air mixture drives the piston back down the cylinder, turning the engine's crankshaft. When the piston completes its compression stroke, an electric spark occurs between the electrodes of the sparkplug. This spark ignites the gaseous mixture, causing it to explode, expand, and thrust the piston to the bottom of the cylinder. Thus, internal combustion actually powers the vehicle: the thermal energy caused by combustion drives the piston downward, and a connecting rod transmits this force to the crankshaft; rotation of the crankshaft creates the mechanical energy that ultimately turns the vehicle's wheels.

Exhaust Stroke

    The exhaust stroke forces spent gases from the cylinder. At the completion of the power stroke, the exhaust port opens. The piston then moves to the top of the cylinder, pushing burnt gases out the exhaust port. When the piston completes its exhaust stroke, the exhaust port closes and the intake port opens, preparing for a fresh four—stroke cycle.

III. CONCLUSION

The sequences of an operation or process are more difficult for readers to combine into a coherent whole than are physical characteristics (Chapter 3). Therefore, you should add a conclusion to the description of action. Such a conclusion takes the readers through a brief, one-paragraph description of a cycle of operation. You do not need details because you have already divided the operation into sequences and thoroughly described it; you must now put it back together, or synthesize it, for the readers. You can accomplish this by tying the topic sentences of each sequence into a general statement of the mechanism's entire operation.

```
                          CONCLUSION

     The piston in a four-cycle engine requires four strokes

(two down and two up) to complete a cycle.  During the intake

stroke, the descending piston draws a fuel-air mixture into

the piston cylinder.  As the piston rises, its compression

stroke forces the mixture against the top of the cylinder.  A

spark from the sparkplug ignites this volatile mixture, and

combustion drives the piston back down the cylinder, turning

the crankshaft, during the power stroke.  As the piston as-

cends again, its exhaust stroke forces the spent gases from

the cylinder.
```

Outline for Describing a Process

As was explained at the beginning of this chapter, process description emphasizes the role of workers instead of the operation of machines. Although a task might require the use of various instruments, a process description concentrates on what the technician does rather than on how the instruments work. For example, a mechanic attempting to set the timing of an automobile needs a timing light; a process description would emphasize the mechanic's actions as he or she manipulates the timing light and other tools to complete the task.

Because both a description of a mechanism in operation and a process description involve action, they are similarly organized and written. Both are divided into sequences, written in the active voice, and require emphasis on causes and effects. Therefore, the body of a process description follows the outline for describing a mechanism in operation. A process description must be introduced somewhat differently, however:

I. Introduction
   A. Definition of Process (also amplification of definition, and general information about process)
   B. Equipment
   C. Major Sequences of Process
II. Description of Process (same as description of operation)
III. Conclusion (same as description of operation)

A sample introduction to the process of soldering follows.

INTRODUCTION

Definition of Process

Soldering forms electrical connections between the components of electrical devices. In this process, soft solder (a tin-and-lead alloy) must be melted over (1) the end of a wire, called the lead, and (2) part of a component, called the terminal. Many such connections are made during the assembly of electrical devices so that current can flow from component to component.

Equipment

The equipment needed for soldering leads to terminals are a soldering iron, a coil of resin-core solder, a knife, and a damp sponge.

Sequences of Process

The major sequences of soldering are preparing the wires and components, tinning the soldering iron, heating the junction, and applying the solder.

The definition-of-process section begins with an operational definition of soldering. The remainder of the paragraph provides general information and familiarizes the reader with the words *solder, lead,* and *terminal,* which will be used throughout the description.

Equipment refers to all mechanisms and materials necessary for performing the process. The heading "Apparatus" is frequently used instead of "Equipment" in laboratory reports, which explain laboratory procedures (or processes). The preceding sample identifies all items needed for soldering, from the soldering iron to the sponge for cleaning it.

The sequences-of-process section then breaks the entire process into manageable units. In the body of the description, each of the sequences will be identified with a heading, and each heading will be followed by one paragraph of detailed description. The sequences must be selected carefully so they logically divide the information to be presented.

The remainder of this chapter completes the description of the process of soldering. Notice that each paragraph has a topic sentence and that the conclusion summarizes the entire process. When writing a process description, resist the temptation to use the imperative (command) voice; your purpose is to *describe* and explain a process rather than to *direct* it. Generally, you should limit the use of the imperative voice to the set of directions provided in manuals, as will be shown in Chapter 13.

PROCESS

Preparing Wires and Components

Prior to soldering, the solderer uses a knife to strip approximately ½ inch of insulation from the end of the wire, creating a lead. Dirt or other matter must then be removed

from both the lead and the terminal. Next, the lead is bent
to form a hook and mounted firmly to the terminal, forming a
junction. If only two wires are being connected, their leads
can be hooked or simply twisted together.

Tinning the Soldering Iron

    The tip of the soldering iron must be kept clean
throughout the soldering process. The solderer accomplishes
this by wiping the tip with a damp sponge, and then applying
a thin layer of solder. This process, called tinning, keeps
the tip clean and provides some protection for the tip as it
heats the junction.

Heating the Junction

    The lead—and—terminal junction must be heated so it will
melt the solder. To accomplish this, the solderer places the
tip of the soldering iron against the junction, simultane—
ously heating both the lead and the terminal. This step and
the next must be accomplished rapidly because excess heat can
damage electrical components.

Applying the Solder

    While still applying heat with the tip, the solderer
places solder against the junction. The solder is held
against the side of the junction opposite the iron because

solder must be melted by the heat of the junction instead of
by directly touching the iron. Otherwise, a "cold" junction might result, causing a weak electrical connection.
After solder flows over the entire junction, the solderer removes the coil of solder, and then removes the iron.

CONCLUSION

The process of soldering electrical components begins
with stripping the wire and attaching it to the terminal of a
component. Then, after tinning the tip of the iron, the
solderer heats the lead-and-terminal junction and allows solder to flow over it. When the junction cools and hardens, it
forms a connection through which current can flow.

The following two student models show the variety of topics that lend
themselves to description of a mechanism in operation and description of
process. The first model describes the operation of a conventional mechanism,
the thermostat; the second model describes the process of developing a roll
of film. Study each of the models so you can point out strengths and weaknesses to other members of your class. Thorough, constructive criticism of
the models will help you write your own description in a way that best
achieves communication.

Models

DESCRIPTION OF A THERMOSTAT IN OPERATION

INTRODUCTION

Purpose of Operation

A thermostat automatically regulates the temperature of a room or an entire building by controlling the furnace. It must be mounted on a centrally located wall so the temperature it senses is representative of the entire area to be heated.

Principle of Operation

The thermostat operates according to the principle that metals expand when heated and contract when cooled. Changes in temperature cause the expansion and contraction of a bimetal (copper and iron) coil within the thermostat. As Fig. 1 shows, copper reacts more than iron to temperature change; therefore, the reaction of the strip of copper determines the bimetal coil's curve.

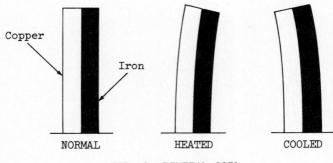

FIG. 1: BIMETAL COIL

Sequences of Operation

    The major sequences of a thermostat's operation are the
on cycle and the off cycle.

OPERATION

The On Cycle

    During the on cycle, the bimetal coil contracts, turning
the furnace on. Figure 2 shows the interior of a thermostat:
the bimetal coil winds around a shaft that rotates, indicating the temperature on a calibrated dial on the thermostat's
exterior (not shown). The outer end of the coil attaches to
a simple mercury switch, an airtight glass capsule that contains 5 grams of mercury. When the bimetal strip becomes
cooler, it contracts, moving counterclockwise. Thus, the
glass capsule moves to the left, causing the mercury to roll
to the left side of the capsule. There the mercury makes
contact with two wires and "shorts" them together; this
closes a circuit connected to the furnace and turns the furnace on.

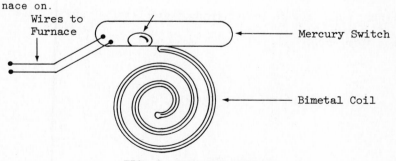

FIG. 2: COIL AND SWITCH

The Off Cycle

The thermostat's bimetal coil expands and turns the furnace off during the off cycle. As the area surrounding the thermostat becomes warmer, the coil (particularly the copper strip) gradually expands. This causes the bimetal coil to become radially larger and to move clockwise. The mercury within the switch therefore rolls to the right, breaking contact with the two wires. The mercury no longer conducts electricity between the two wires, so the circuit to the furnace breaks, turning the furnace off.

CONCLUSION

A thermostat operates on the principle that metals contract when cooled and expand when heated. The contraction of a bimetal coil within a thermostat activates a mercury switch connected to the furnace, turning the furnace on. When the bimetal coil expands, however, it causes the mercury within the switch to break contact with the two wires. As contact breaks, the wires no longer conduct electricity, and the furnace turns off.

DEVELOPING A ROLL OF FILM

INTRODUCTION

Definition of Process

Developing a roll of film produces a negative of the image captured by the camera. The process should not be confused with printing, which creates a photograph from a negative. This report will describe only the process of developing black—and—white film.

Equipment

Processing a roll of film requires a dark room, a film tank, a timer, a supply of water, and bottles of the following three chemicals: developer, stop bath, and fixer. A funnel can also be helpful for pouring chemicals back and forth betewen the tank and bottles.

Sequences of Process

The six sequences in developing film are loading the film into the tank, developing, stopping development, fixing the film, washing, and drying.

PROCESS

Loading Film into the Tank

Film must be processed in a tank that allows liquids to enter and leave without letting light in. In darkness, the technician removes the film from its case and then wraps the film around a film-tank reel, taking care not to touch the film. The technician then places the reel in the tank and closes the lid. With the film safely in the light-tight tank, the technician can complete the remaining sequences in normal room light.

Developing

Developing darkens areas on the film that were exposed to light. A chemical called developer must first be mixed with water and then poured into the film tank. Specifications supplied by the manufacturers of the chemical and the film must be followed very carefully: the chemical manufacturer specifies the amount and temperature of water used to dilute the chemical, and the film manufacturer provides a list of development times necessary for various water temperatures. After pouring the solution into the tank, the technician firmly taps the tank several times to release any air bubbles formed on the film surface. The chemicals must then be agitated periodically to distribute them evenly. The technician does this initially for 15 seconds by inverting the tank several times, and repeats the agitation for 5 seconds every 30 seconds after that. During the last 10 seconds of this sequence, the technician drains the developer from the tank.

Stopping Development

Because the surface of film is somewhat porous, a certain amount of developer remains on the film and continues
development after draining. The technician therefore immediately pours a chemical called stop bath into the tank to neu-
tralize the developer. The stop bath should be the same temperature as the developer. During the next 30 seconds, the
technician inverts the tank several times to agitate the
chemicals and then drains the tank again.

Fixing the Film

Fixing removes all the undarkened chemicals from the
film, leaving dark areas that correspond to the light areas
of the image. Fixer must be mixed according to the manufacturer's specifications and used at the same temperature as
the developer and stop bath. The technician adds fixer as
soon as the stop bath is drained from the tank and then taps
and agitates the tank, as was done during the development sequence. At the end of the time specified by the film manufacturer, the technician drains the tank. Once fixing is
completed, the film is fully developed and needs only to be
washed and dried.

Washing

The film must be thoroughly washed to remove all resid-
ual chemicals. This is done most effectively with running
water. Film-tank manufacturers usually make an attachment
that inserts into the opening in the lid and forces water
across the film; however, the technician may simply open the
lid and run water into the tank. The film must be washed for
an hour, and the temperature of the wash should be between 55
and 75 degrees to avoid damage from cracking or melting.

Drying

After the film is washed, it must be dried in a dust-
free area before it can be printed. The technician carefully
removes the film from the reel, clips one end to a clothes-
line, and puts a clip on the other end to hold it down. The
film can then air dry. If the film is needed immediately, it
may be hung inside one of the commercially available film
dryers. Once the film is dry, the technician cuts it into
strips and stores the strips in envelopes until printing
them.

CONCLUSION

Developing a roll of black—and—white film starts with loading the film into the tank. The technician adds developer to darken the exposed areas on the negative and then stops development by using stop bath. All undeveloped chemicals dissolve during fixing. Finally, the technician washes and dries the film. The result is a transparent negative reproduction of the image captured by the camera.

Writing Assignment

Write a 500-word description of a mechanism in operation or description of a process for an uninformed reader. The listed topics simply suggest the wide variety of subjects possible. Search your own experience to think of mechanisms you are familiar with and processes you have performed. Topics are available from your specialized technical field and from your experience on the job, at the workbench, in the kitchen, or in the laboratory.

Operations	*Processes*
solar furnace	installing ceramic tile
carburetor	repairing a dent in a car
compass	tuckpointing a chimney
air conditioner	antiquing a table
human heart	installing an electrical outlet

When writing the description, keep in mind that your objective is to *describe* an operation or process; avoid writing a set of directions. Feel free to use visuals wherever they reinforce your description. Number and title each visual, and refer to it at an appropriate place in your paragraphs of description.

Exercises

1. Using the drawings shown, write a one-paragraph description of a paper clip, clarifying its size, shape, and dimensions. Then devote a paragraph or two to detailed description of how a paper clip operates. You will have to explain torsion, the principle by which a clip works, so check a dictionary before writing. Aim your paragraphs at a technically uninformed reader.

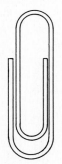

2. In one paragraph, describe what happens as a person sets up a folding chair. The topic sentence of your paragraph should explain the purpose of this process, and the remaining sentences should provide details. *Describe* the process rather than giving commands, and use the word *operator* rather than *you*. Observing the setting up of an actual folding chair might be helpful to you.

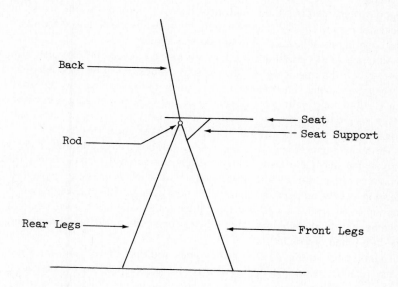

5 Interpreting Statistics

THE interpretation of statistical data is the process of deriving and communicating logical conclusions from a body of facts. A widely held misconception is that data, or facts, somehow speak for themselves. If they do, they speak with ambiguity because two people examining the same data often arrive at totally different conclusions. Until the report writer interprets them, statistical data are nothing but ink marks on pieces of paper. Through interpretation, the facts do speak, and if the writer has been rigorously logical, they speak with authority.

Although most people are not fully aware of it, everyone interprets data every day. When deciding whether to pass a truck, the driver considers the acceleration of the car, the speed of the truck and oncoming cars, the road, the visibility, and perhaps how late he or she is before making a decision. Many everyday choices, some more crucial than others, require such interpretation, and so do the daily decisions in business and industry.

Those in management and supervisory positions are constantly making moves that affect the profit and ultimately the survival of their firms. The problems range from critical matters—such as whether or not to expand, to develop a new product, or to install a new system—all the way down to routine situations—such as promoting the most qualified employee, deciding whether a welded or soldered joint is better for a certain structure, or selecting the best brand of fork lift.

Interpreting statistics is one of four steps you take as you prepare a recommendation report. Although the steps overlap each other, the process of writing a recommendation report always involves the collection, analysis, presentation, and finally, interpretation of data. Depending on the complexity of the data, this process can require simply asking a few questions and writing a short recommendation report, or spending months sifting through data and compiling a formal feasibility report.

Collection of Data

Generally speaking, the more information examined, the greater the probability of making the right recommendation. Four methods of data collection are observation, testing, interviewing, and studying printed material.

Direct observation is probably the most common way of getting information. For example, to figure out when a shipment will be ready for a customer, a writer observes and measures the time it takes to manufacture a certain number of parts; he or she then uses that data to estimate the completion time for the entire order. As in all methods of information gathering, the observer has to insure that the data are accurate, consider the possibility of equipment breakdowns, and estimate the potential effect of such breakdowns on the production rate.

Testing is similar to observation, except that the conditions are more tightly controlled. For example, the strength of various types of steel can be tested to determine the best type for a particular bridge. The testing occurs in a laboratory, where technicians create conditions identical to those the steel must withstand when the bridge is in use.

Conducting interviews is another effective method of gaining information. Anything from carrying on an informal discussion with a fellow employee to interviewing a local authority may prove helpful. A writer who must recommend the best brand of equipment can request opinions from other firms. Such interviews and surveys do not always yield statistical data, but they may provide information leading to the correct recommendation.

Printed material serves as an excellent source of data for recommendation reports. Technical journals contain information about contemporary trends and innovations, precisely the type of data a writer needs to make recommendations affecting the future of the department or firm. Sales literature from manufacturers is also available, and many firms have libraries on technical subjects. Chapter 6 will provide detailed information about collecting and using printed material.

Analysis of Data

The only effective way to make data meaningful is to analyze them, or separate them, according to criteria. The selection of logical criteria is crucial to the entire recommendation report, because criteria are the bases on which the recommendation will ultimately be made. To explain what criteria are and how they are selected, let's create a hypothetical problem situation.

Suppose that you have been assigned to write a report recommending the best kind of water heater to install in a small commercial building. Three kinds of water heaters are available: natural gas, oil, and electric.

You have attempted to identify important aspects of the heater problem and to gather data on the three heaters from a local commercial heating company. You must now decide how to divide the data logically according to criteria. You carefully examine the data and identify six tentative criteria:

1. cost of water heaters
2. cost of installing water heaters
3. cost of water heaters in operation
4. cost of fuels
5. availability of fuels
6. ability of water heaters to supply sufficient hot water

Having selected six tentative criteria, you must now make sure that sufficient data are available and that the criteria do not overlap or repeat each another. Such repetition would confuse or mislead the readers of your report. You should also see if you can logically combine any of the criteria.

First, you decide that the first two tentative criteria, cost of water heaters and cost of installing water heaters, can be combined into a single criterion, initial cost. Further study of the data shows you that the cost of operating each heater largely depends on the cost of the fuel (gas, oil, electricity) the heater uses. Therefore, you can also combine the second two tentative criteria, cost of water heaters in operation and cost of fuels, into a single criterion, operational cost.

Then, your investigation shows that although the three fuels are now available, information about their future supply is vague and unreliable. You must therefore reject the availability-of-fuels criterion, and you must inform your readers in the introduction to your report that reliable data are not available. Finally, you learn from your study of water heaters that you should change the name of your fifth criterion, ability of water heaters to supply sufficient hot water, to recovery rate. Thus, your examination of the criteria and the data has reduced your original five criteria to three:

1. cost of water heaters   } *Initial*
2. cost of installing water heaters } *cost*
3. cost of water heaters in operation } *Operational*
4. cost of fuels   } *cost*
5. ability of water heaters to supply } *Recovery*
   sufficient hot water   } *rate*

A report writer must always carefully consider all aspects of a problem and thoroughly analyze data in order to select appropriate criteria. In the preceding example, failure to identify operational cost as a criterion would have led to a recommendation of the wrong water heater. The careful selection of criteria also forces a writer to reject data that do not really apply to the problem situation. Some writers want to put all the data they have collected into the report to show how hard they have worked. The report readers, of course, will remember only the time they wasted reading meaningless data.

Presentation of Data

When presenting statistical data in a report, you must decide whether to fit them into paragraphs or place them in a table. In making this decision, remember that words and numbers are very different kinds of symbols that communicate best in different environments. Words obviously communicate well in an environment of sentences and paragraphs, but numbers are communicated better in tables. The following paragraph, presenting costs of water heaters, fails to communicate well because it forces the readers to create a table in their minds.

A heater's initial cost is the total of its purchase price and installation cost. The purchase price of an oil heater is $290, which is greater than the others because it must include the purchase of an oil-storage tank. The gas and electric heaters have purchase prices of $205 and $250, respectively. The gas and oil heaters each cost $90 to install, compared with $50 for the electric heater. Installation of gas and oil heaters costs more because both of them must be vented to the outside of the building. Overall, the initial cost of the heaters is $295 for gas, $380 for oil, and $300 for electric.

Presented within a paragraph, the statistical comparisons are unnecessarily difficult to grasp. On the other hand, a brief look at the following table enables the readers to understand the information easily. Organizing statistical data into tabular form pays off in communication.

TABLE 1: INITIAL COST

	Gas	Oil	Electric
Purchase	$205	$290	$250
Installation	90	90	50
TOTAL	$295	$380	$300

Interpretation of Data

As was stated at the beginning of this chapter, a smart writer assumes responsibility for interpreting the data presented in a report. The interpretation
of data, or telling the readers what the data mean, avoids the dangerous
practice of hoping that your readers will get everything out of the data that
you intend.

The following paragraph, which interprets the data in Table 1, exemplifies
two main kinds of interpretation. First, it explains *reasons* for important differences in initial cost, namely, that an oil heater requires the purchase of an
oil-storage tank and that the oil and gas heaters must be vented to the outside.
Second, it *emphasizes* the $80 difference in total cost. Although this infor-
mation is evident in the table, stating it in the paragraph does not constitute
repetition; the paragraph simply emphasizes particularly significant data that
the writer wants the readers to remember.

A heater's initial cost is the total of its purchase
price and installation cost (Table 1). The purchase price of
an oil heater is greater than the others because it must include the purchase of an oil-storage tank. The oil and gas
systems are more costly to install than the electric system;
both of them must be vented to the outside of the building.
Overall, the oil system costs about $80 more than the electric
and gas systems.

Keep in mind that the data just interpreted are fairly simple, and that interpretation becomes increasingly important as the data become more complex. Notice also that the paragraph refers its readers to the table. As was explained in Chapter 7, you must always refer to tables and figures in the paragraphs of a report. In order to control your readers' assimilation of a report, you must tell them when to examine a table, and then state what the table means. You can do this if the table appears after the paragraph of interpretation, but not if the table precedes its interpretation.

Outline for the Interpretation of Statistics

The principles for the interpretation of data are applicable to both informal recommendation reports and formal feasibility reports. Feasibility reports (to be discussed in Chapter 11) are reserved for problems requiring the interpretation of extremely complex data. Recommendation reports, which are more common and much shorter, can be organized according to the following outline:

I. Introduction
   A. Purpose of Report
   B. Definition of Problem
   C. Method of Investigation
   D. Scope (alternatives and criteria)
II. Presentation and Interpretation of Data
   A. Judgment According to First Criterion
   1. Definition of Criterion
   2. Presentation of Data for All Alternatives
   3. Interpretation of Data
   B, C, etc.: Other Criteria (same as for First Criterion)
III. Conclusions
IV. Recommendation

I. INTRODUCTION

Although an introduction should be concise, it must prepare the reader for information presented in the body of your report. This requires an identification of the purpose of the report, a definition of the problem your report attacks, and a statement of the scope of the report.

A. Purpose of Report   Begin a recommendation report with a straightforward statement such as "The purpose of this report is. . . ." You can generally cover the purpose, which is to recommend a solution to a particular problem, in one sentence.

B. Definition of Problem   Stating the purpose of the report leads you directly into a definition of the problem, which is basically an explanation of what has made the report necessary. Here you must remember that as the person who has investigated the problem, you are more knowledgeable than your readers are. You might have found that the problem is more complex than it first seemed to you and the person who assigned it to you. Possibly your investigation has shown that a particular aspect of the problem requires special emphasis. Readers unaware of this background information might not fully understand statements in the body of your report, and thus they might not accept your recommendation.

C. Method of Investigation   After defining the problem, you must state your method of gathering information to solve the problem. The four major methods of gathering data (described earlier in this chapter) are observing, testing, interviewing, and reading printed material. Stating the source of your data not only gives credit where it is due but lends authority to your data and thus to your report.
   In the introduction, a general statement of your method of investigation is generally sufficient: "Data for this report were gathered from sales literature supplied by IBM Corporation and Data 100 Corporation." If, however, the body of your report contains a table taken directly from a printed source, a source statement should appear immediately below the table, as will be exemplified in Chapter 7.

D. Scope   Having identified your method of investigation, your next logical step is to identify the alternatives and the criteria you used to judge the data. You should simply name the alternatives, or possible solutions to the problem, in the introduction. The criteria do not require elaboration in the introduction because you will define them individually in the body of the report. In the section on scope, you should name them in the order they appear in your report. If you have not included a criterion in your report because data are unavailable or unreliable, you should state this in the section on scope so your readers will not think that you have overlooked the criterion.

INTRODUCTION

Purpose

The purpose of this report is to recommend a water heater for a commercial building.

Problem

Superior Construction Company is contracted to construct a small commercial building and needs to know what type of water heater to install. The company has determined that the heater must be able to heat 80 gallons of water per hour to adequately supply the building.

Method of Investigation

The information in this report was gathered during an interview with Mr. David Tomlinson of Wray Commercial Heating Company, Lake Charles, Louisiana, March 29, 1977.

Scope

The three types of water heater available are natural gas, oil, and electric. At present, these fuels are available, but this report does not consider their future supply because reliable data are unavailable. Each of these heaters will be evaluated according to its recovery rate, initial

cost, and operational cost.

II. PRESENTATION AND INTERPRETATION OF DATA

Each criterion mentioned in your section on scope becomes a main heading in the body of the report, as shown in the sample that follows. A single paragraph is usually sufficient for a criterion, and as the outline specifies, you begin the paragraph with a definition of the criterion. The reader needs to know what is meant by *recovery time* and by *initial cost*.

The remainder of the paragraph is usually devoted to interpreting data you have presented in a table. The method for doing this (explained earlier in the chapter) is exemplified by the second and third paragraphs in the following report. Sometimes, however, the data for a criterion are so few that they do not require a table. In such cases, you may both present and interpret the data in a paragraph, as the first of the following paragraphs shows.

RECOVERY RATE

The recovery rate of a water heater is a measure of the

gallons of water it can heat to 100 degrees Fahrenheit in an

hour. The gas and oil heaters have recovery rates of 150

gallons per hour, and the electric heater's rate is 100 gal-

lons per hour. Therefore, each of the heaters meets the 80-

gallons-per-hour requirement of the contractor.

INITIAL COST

A heater's initial cost is the total of its purchase

price and installation cost (Table 1). The purchase price of

an oil heater is greater than the others because it must in-

clude the purchase of an oil-storage tank. The oil and gas

systems are more costly to install than the electric system;
both of them must be vented to the outside of the building.
Overall, the oil system costs about $80 more than the electric
and gas systems.

OPERATIONAL COST

The operational cost of a water heater is the average
cost of heating 100 gallons of water to 100 degrees Fahrenheit. This cost depends on (1) the cost of fuel, and (2) the
heater's efficiency in converting the fuel into heat. Although an electric heater is considered 100% efficient in
converting electricity into heat, the high price of electricity (Table 2) makes its yearly operational cost $659 per year.

TABLE 1: INITIAL COST

	Gas	Oil	Electric
Purchase	$205	$290	$250
Installation	90	90	50
TOTAL	$295	$380	$300

TABLE 2: OPERATIONAL COST

	Gas	Oil	Electric
Cost of heating 100 gal to 100 deg	$.31	$.25	$.33
Gal per day	640	640	640
TOTAL COST (YEAR)	$619	$499	$659

Gas and oil heaters are only 70% efficient as converters because heat escapes through ventilation pipes. Gas and oil,
however, currently cost less than electricity; the yearly operational cost is $619 for a gas heater and $499 for an oil
heater.

III. CONCLUSIONS

After presenting and interpreting data for all the criteria, you are ready to draw conclusions. Your conclusions section should summarize the most significant information for each criterion covered in the report. One sentence about each criterion is usually sufficient to prepare the reader for your recommendation.

CONCLUSIONS

All three heating systems can supply the commercial
building with sufficient hot water. The natural gas and
electric heaters each cost about $300 initially, but each also
costs more than $600 per year to operate. Although the oil
heater costs more initially, $380, it costs only $499 per year
to operate.

IV. RECOMMENDATION

If your recommendation surprises the reader, your report fails. Carefully written reports build steadily toward the recommendation, often reducing it to a formality. For short reports like the sample presented here, one or two sentences should suffice. For complex reports that involve many aspects of a problem, a longer paragraph may be necessary.

RECOMMENDATION

The oil heating system is recommended because within about 12 months, its lower operational cost will compensate for its higher initial cost.

Dangers with Statistical Data

1. Be careful of bias. After making sure the sources of the data are good, let the data go to work for you in your effort to make the right decision. If you happen to make the wrong recommendation based on careful interpretation of data, you at least have the data to back you up. If you base a wrong recommendation on your own preconceptions, you have nothing.
2. Make sure that evidence you use to support your recommendation is valid. The fact that a particular system works fine for company X does not necessarily mean the system will work for your firm. Perhaps company X is larger, has a lower caliber of employees, and emphasizes quantity rather than quality. These things make a difference.
3. Do not assume that because alternative A is bad, alternative B must be good. Maybe both are bad. Also, do not assume that you should recommend A simply because A is good. Maybe B is better. Examine each of the alternatives thoroughly and objectively.

Two student-written reports follow, the first evaluating voltmeters and the second comparing gasoline and diesel truck engines. Read the models carefully and be prepared to discuss their strengths and weaknesses.

Is the first report understandable by those of you not majoring in electronics? Is the second report clear to those with little automotive knowledge? These are important questions, because in industry, recommendations involving the spending of money must often be approved by people specializing in areas of management rather than technology.

Do the paragraphs of both reports adequately define the criteria and interpret the data presented in the tables? Would you have selected different or additional criteria for the reports? Do the reports contain sentences that would be clearer if they were organized differently?

Models

RECOMMENDATION OF A VOLTMETER

INTRODUCTION

Purpose

The purpose of this report is to recommend a voltmeter for a university's electronics laboratories.

Problem

The School of Basic and Applied Science at Parker University has decided to replace the meters in some of its electronics labs. The meter must be capable of measuring dc voltage, dc current, ac voltage, ac current, and resistance.

Method of Investigation

Data for this report were gathered from sales literature supplied by Precision Instruments Corporation, Memphis, Tennessee.

Scope

The university has a choice between two meters, a digital-readout meter or an analog meter (one that uses a needle

to indicate voltage). The meters will be evaluated in terms
of durability, accuracy, sensitivity, and cost.

DURABILITY

Durability refers to the ability to resist wear and damage during use. The analog meter has moving parts subject to
wear, and its needle can become bent, resulting in false
readings. This meter can also be easily damaged if it is im-
properly connected to the circuit for making measurements.
The digital meter has no moving parts subject to wear, and it
also has internal protective circuits that prevent damage
from improper connections. Therefore, the digital meter will
withstand abuse better than the analog meter.

ACCURACY

Accuracy refers to the precision of a meter's measurements. False readings can confuse an important experiment or
hinder the students' understanding of a concept. Table 1
shows the two meter's accuracy in performing various functions. The digital meter is more accurate for each of the
functions.

SENSITIVITY

The sensitivity of a meter refers to how small a reading it can take. A more sensitive meter can be used in a wider variety of experiments. Table 2 shows the two meter's sensitivity in measuring different functions. Although their sensitivity varies according to what is being measured, both meters have adequate sensitivity for university laboratory use.

TABLE 1: ACCURACY

Function	Digital	Analog
dc voltage	±0.0003[a]	±0.02
dc current	±0.0011	±0.02
ac voltage	±0.0005	±0.015
ac current	±0.0005	±0.02
Resistance	±0.0002	±0.05

[a] ± means "plus or minus"

TABLE 2: SENSITIVITY

Function	Digital	Analog
dc voltage	0.01 volt	0.01 volt
dc current	0.0001 amp	0.001 amp
ac voltage	0.1 volt	0.01 volt
ac current	0.0001 amp	0.001 amp
Resistance	100 ohms	10 ohms

COST

The digital meter costs $500, and the analog meter costs $445. The digital meter costs more because of the complex circuitry necessary to display readings in the form of numerals.

CONCLUSION

The digital and analog meters both have adequate sensitivity, but the digital meter is less likely to be damaged during use, and it is more accurate. The digital meter, however, costs $55 more than the analog meter.

RECOMMENDATION

The digital meter is recommended despite its higher cost. Its durability and accuracy make it ideal for student laboratory use.

RECOMMENDATION OF A TRUCK TRACTOR

INTRODUCTION

Purpose

The purpose of this report is to recommend a truck tractor for a commercial tank-line company.

Problem

The Judson Tank-Line Company has a contract with American Chemical Corporation to transport liquid chemicals between an eastern seaboard plant and a midwestern plant. The
tank-line company has specified that the tractor must be able
to pull a load of 30,000 pounds at 50 miles per hour in order
to keep the number of trips between plants at a minimum.

Scope

The two kinds of truck tractors evaluated are gasoline
powered and diesel powered. These alternatives will be examined according to their payload, initial cost, and operating
cost.

Method of Investigation

Data for this report were gathered from sales literature supplied by Ford Motor Company and Mack Truck Company.

PAYLOAD

The payload of a truck tractor is the weight that it can pull at a specified speed. The gasoline-powered tractor has a payload of 35,000 pounds at 55 miles per hour, and the diesel tractor has a payload of 45,000 pounds at 55 miles per hour. Therefore, both tractors meet the company's requirement of 30,000 pounds at 50 miles per hour.

INITIAL COST

A tractor's initial cost is the total of its purchase price and vehicle license. As Table 1 shows, the purchase price of the diesel tractor is greater than that of the gasoline tractor; diesel-engine components are more expensive

TABLE 1: INITIAL COST

Item	Gasoline	Diesel
Purchase	$28,500	$34,100
Vehicle license	2,000	1,500
TOTAL	$30,500	$35,600

than gasoline–engine components. These tractors must travel from the eastern seaboard to the midwest, and the cost of licenses depends on the pollutants they emit into the air. Because a diesel engine emits fewer pollutants than a gasoline engine, its license costs less. Overall, the diesel tractor costs $5100 more than the gasoline tractor.

OPERATIONAL COST

The operational cost of a truck tractor is dependent on the cost of fuel and the tractor's efficiency in converting fuel into useful work (miles per gallon). The gasoline tractor's operational cost is greater than the diesel's (Table 2) because gasoline costs more and gets fewer miles per gallon than diesel fuel. Overall, operation of the diesel costs $3300 less per year than operation of the gasoline tractor.

TABLE 2: OPERATIONAL COST

Item	Gasoline	Diesel
Mi per gal	5.0	5.5
Fuel cost (per gal)	$ .62	$ .50
TOTAL (per 100,000 mi)	$12,400	$9,100

CONCLUSIONS

Both kinds of tractors can pull the load specified by
the tank company. The diesel-powered tractor initially costs
$5100 more than the gasoline-powered tractor; however, the
diesel tractor costs $3300 less per year to operate.

RECOMMENDATION

The diesel-powered tractor is recommended because within
18 months, its higher initial cost will be offset by its
lower operational cost.

Writing Assignment

Assume that you are working for a local firm and have been asked to evaluate two kinds, brands, or models of equipment. Select a limited topic (not Ford versus Chevrolet or the like) and evaluate the alternatives in detail. Write a 400-word report recommending that one of the alternatives be purchased to solve a problem. Both alternatives should be workable; your report must recommend the one that will work better.

Gather data about the alternatives just as you will when working in industry: from sales literature, dealers, your own experience, and the experience of others who have worked with the equipment. Select a maximum of four criteria by which to judge the alternatives, and use a minimum of one table in the report. Aim your report at someone not familiar with the equipment.

Exercises

1. Statistical data often become needlessly confusing when they are presented within a paragraph. Sort out the statistics in the following paragraph and place them in their more natural environment, a table.

```
      The turntable manufactured by ABC Company costs $119.95,

versus $135.50 for a comparable XYZ unit.  The base (cabinet)

for ABC's turntable is priced at $7.50, and the cartridge

sells for $39.95.  XYZ Company's cartridge costs $37.50, and

the cabinet is priced at $10.00.  The companies charge the

same amount, $7.50, for plastic dust covers.  The entire ABC

turntable assembly costs a total of $174.90, while an XYZ unit

costs a total of $190.50.
```

2. Which team won? Statistics are notorious liars, but careful analysis and interpretation of the following data from a college football game should enable you to reach a logical conclusion. Start by narrowing the number of criteria by which to base your decision. For example, you can reject the yards-penalized criterion because the teams' statistics are not significantly

different. Do not be confused by the passing statistics: team X attempted 49 passes, completed 29, and had 3 intercepted. In analyzing the data, you may decide to create a main criterion of your own and to make subcriteria of some of the existing criteria.

Write your interpretation in paragraph form, and define potentially confusing football terminology. During class discussion after completion of this exercise, defend your decision.

	Team X	Team Y
First downs	26	26
Yards rushing	70	396
Yards passing	308	74
Return yardage	3	37
Passes attempted	49	19
completed	29	9
intercepted	3	1
Punts	6	4
average	35	42
Fumbles lost	1	0
Yards penalized	72	70

3. Assume that a friend of yours, someone interested in pursuing a bachelor's degree in your technical field, has asked you for advice. Your friend is trying to decide whether to enroll at a four-year school several hundred miles from home, or whether to attend a community college for two years before transferring to the four-year school.

To assist him or her, evaluate the alternatives (using real schools) and write a recommendation report. Your first task is to select logical criteria on which to base the recommendation. The criteria may include the total cost of attending each school, the availiability of part-time jobs, the transfer of credits to the four-year school, or any others that you think are important.

Gather data for each of the alternatives and structure the report according to criteria. All the data need not be statistical but each of your ideas must be supported by facts. To insure that your friend clearly understands the information, consider presenting complex data in visual form.

6 Researching Published Information

FOR technical students, the technique of researching published information has value beyond the writing of acceptable term papers. Technologies produce volumes of new material daily, making knowledge of the library useful for experienced professionals as well as students. Published information helps keep industry informed about up-to-date methods, equipment, and products. That is why technical associations publish journals. It also explains an important function of library research in technical curricula: to provide one of the skills necessary for self-education.

This chapter discusses the basic techniques for locating and using published information. In many ways, library research resembles laboratory research. Successful laboratory researchers know their equipment, devise plans for reaching their objectives, and keep careful records. This systematic, scientific approach, which is familiar to all technical students, works just as effectively in library research.

Where to Start

On entering the library, you confront thousands of books and periodicals, knowing that perhaps 10 of them are potentially good sources.

The following paragraphs explain how to locate these 10 sources. Encyclopedias may be useful if they refer you to pertinent books and periodicals. Generally, however, you will immediately examine the card catalog and periodical indexes to locate potentially useful publications.

ENCYCLOPEDIAS

Encyclopedias serve only as points of departure because they do not provide the type of detailed information needed in reports. They do give background information, however, and many of their articles contain bibliographies naming sources of more specific and useful data. The four most popular encyclopedias are the following:

Encyclopedia Americana: considered a good source for general technical information; supplemented yearly by the *American Annual*
Encyclopaedia Britannica: considered the best general encyclopedia; supplemented annually by the *Britannica Book of the Year*

Encyclopedia of Chemical Technology: covers all fields of chemical technology; 15 volumes; now includes 2 supplemental volumes
Encyclopedia of Science and Technology: covers major technological applications of all the natural sciences; 15 volumes; supplemented yearly

The following are smaller, specialized encyclopedias:

The Encyclopedia of Chemistry
The Encyclopedia of Chemical Process Equipment
The Encyclopedia of Electronics
The Encyclopedia of Management
The Encyclopedia of Physics

THE CARD CATALOG

The card catalog lists every book in the library. Many libraries list each book in three alphabetical files containing (1) author cards, (2) title cards, and (3) subject cards, as shown in Fig. 6.1. Except for their main headings, the cards are exactly alike. Items on the subject card in Fig. 6.1 are numbered for purposes of reference.

1. You begin by thinking of subjects under which you might find pertinent information. Then, examine the cards in the subject section of the card catalog. In Fig. 6.1, the subject is population. Subject cards are heavily cross-referenced, so they often refer you to another subject. Also, some books are listed under several subjects, as will be seen in item 8.
2. The call number explains where the book can be found on the shelves. If the call number begins with letters, as in Fig. 6.1, the library is organized according to the Library of Congress system. Where the Dewey system is used, this book's call number is 301.3, as specified at the bottom of the card.
3. The author's name appears here. To see if he or she has written other books that might be useful, check the author cards.
4. The name of the book appears below the author's name.
5. The place of publication, publisher, and year of publication are listed in that order. The year is important if rapid advancements are being made in your subject area.

6. The collation gives the total pages of the book and, if applicable, indicates that it has illustrations or is part of a series. Other items, of little interest, might be the price and the height of the book in centimeters.

7. This section provides special information about the contents of a book, such as its inclusion of a bibliography. Bibliographies are extremely helpful because they lead you to additional sources of information.

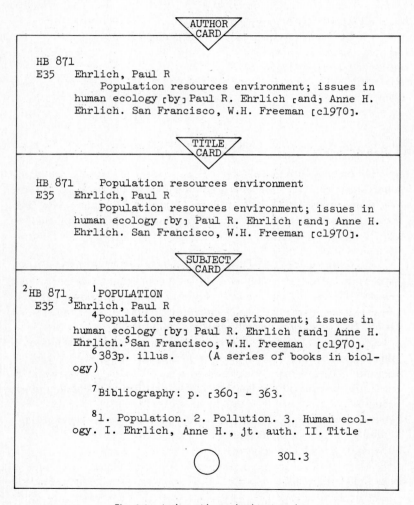

AUTHOR CARD

HB 871
E35   Ehrlich, Paul R
           Population resources environment; issues in
       human ecology ₍by₎ Paul R. Ehrlich ₍and₎ Anne H.
       Ehrlich. San Francisco, W.H. Freeman ₍c1970₎.

TITLE CARD

HB 871   Population resources environment
E35   Ehrlich, Paul R
           Population resources environment; issues in
       human ecology ₍by₎ Paul R. Ehrlich ₍and₎ Anne H.
       Ehrlich. San Francisco, W.H. Freeman ₍c1970₎.

SUBJECT CARD

[2]HB 871   [1]POPULATION
E35   [3]Ehrlich, Paul R
           [4]Population resources environment; issues in
       human ecology ₍by₎ Paul R. Ehrlich ₍and₎ Anne H.
       Ehrlich.[5]San Francisco, W.H. Freeman ₍c1970₎.
           [6]383p. illus.   (A series of books in biol-
       ogy)

           [7]Bibliography: p. ₍360₎ - 363.

           [8]1. Population. 2. Pollution. 3. Human ecol-
       ogy. I. Ehrlich, Anne H., jt. auth. II. Title

                                         301.3

Fig. 6.1   Author, title, and subject cards

8. Tracings name other subjects in the card catalog that list the book. Look under those subjects to insure that you cover all possible sources of information.

PERIODICAL INDEXES

Periodical indexes help you find magazine articles in the same way that card catalogs assist in locating books. Periodicals are published for all technical areas, and as was mentioned earlier, they are especially good sources for keeping you informed of innovations in your field. Four major periodical indexes for technical subjects are the following:

Applied Science and Technology: subject index to articles in technology, engineering, science, and trade, including the areas of automation, construction, electronics, materials, telecommunication and others
Business Periodicals Index: subject index to articles in all areas of business, including automation, labor, management, finance, marketing, public relations, communication, and others
Engineering Index: subject-author index to publications of industrial organizations, government, research institutes, and engineering organizations; includes annotations (descriptions) of the articles
Reader's Guide to Periodical Literature: subject-author index to articles on general subjects

A list of articles appears beneath each subject heading in an index, as shown in Fig. 6.2, an excerpt from the *Applied Science and Technology Index.* Subheadings appear beneath many subject headings to identify more specific areas of information. The subheadings beneath "Air Pollution," not shown in Fig. 6.2, include "Air Pollution Laws and Regulations," "Research," and "Testing."

Figure 6.3 deciphers the first entry in Fig. 6.2, an article on pollution by Dermot O'Sullivan. With this information, you need only the magazine's call number, found in the periodicals section of the card catalog, to complete a bibliography card for the source.

BIBLIOGRAPHY CARDS

As you find potential sources of information in the card catalog and periodical indexes, list them on separate 3-by-5-inch cards. These bibliography cards should contain the name of the author, the title of the article or book, and facts about the book's publication. You will use this information when documenting the sources of information used in the report. For your immediate use, write down the call number and any special information about the source. Figures 6.4 and 6.5 are sample bibliography cards.

```
AIR pollution


Air pollution. D.A. O'Sullivan. il Chem & Eng N 48:38-41+
   Je 8 '70 (reprints 50¢)

Air pollution and human health. L.B. Lave and E.P. Seskin,
   Science 169:723-33 bibliog (p 731-3) Ag 21 '70


Air pollution by motor exhaust and other combustion pro-
   ducts. B. Szczeniowski. bibliog Eng J 53:4-10 Jl '70


Air pollution consultant guide/1970. map Air Pollution
   Control Assn J 20:495-9 Jl '70
```

Fig. 6.2   Excerpt from a periodical index

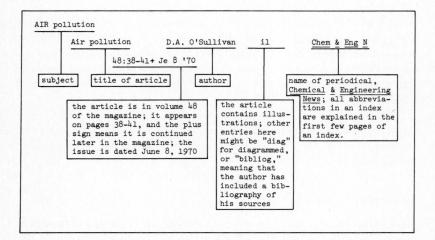

Fig. 6.3   Periodical entry explained

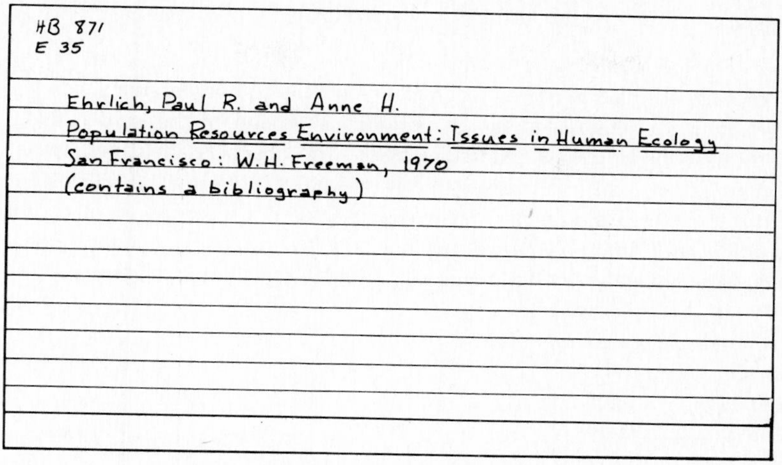

HB 871
E 35

Ehrlich, Paul R. and Anne H.
Population Resources Environment: Issues in Human Ecology
San Francisco: W.H. Freeman, 1970
(contains a bibliography)

Fig. 6.4   Bibliography card for book

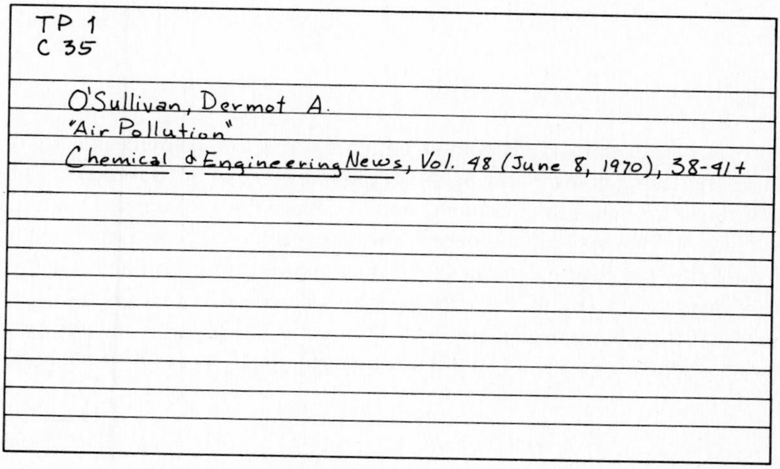

TP 1
C 35

O'Sullivan, Dermot A.
"Air Pollution"
Chemical & Engineering News, Vol. 48 (June 8, 1970), 38-41+

Fig. 6.5   Bibliography card for periodical

ABSTRACTS

After completing bibliography cards for all the sources, you can examine the appropriate abstracts to narrow your list of potentially good periodical articles. Abstracting journals, listed in the periodicals section of the card catalog, briefly describe articles in particular fields. The abstracts are generally descriptive rather than informative; that is, they state the scope of articles but do not summarize the articles. Therefore, they do not serve as substitutes for the article and cannot be quoted, but they do allow you to reject inappropriate articles immediately. The following is a brief list of technical abstracts:

Aeronautical Engineering Index
Chemical Abstracts
Engineering Abstracts
Geological Abstracts
Metallurgical Abstracts
Mineralogical Abstracts
Nuclear Science Abstracts
Science Abstracts (divided into sections on electrical engineering and physics)

Taking Notes

With your bibliography cards completed, go directly to the sources to begin reading and taking notes. Knowing the purpose and scope of your report, you already have a rough idea of its main headings. Place the appropriate heading, the author of the work, and the page number on each note card. Even though you may alter the headings somewhat during your research, some uniformity will be maintained among the notes. Each card should contain notes on a single subject and from a single source. This seems like a waste of cards, but it greatly simplifies your card shuffling when you finally organize the report.

Quoting and Paraphrasing

Quoting, or copying the words of the author verbatim, is done when a passage contains precisely the information needed for the report. An original passage from an article in *Chemical & Engineering News* is shown in Fig. 6.6. When you quote from such a passage, put quotation marks around each quote and

Cleanup

"Technology is responsible in large measure for pollution, so let's use technology to clean it up," is the stand that many enviromentalists take. They also frequently cite the fact that if we can land men on the moon and bring them safely back to earth again, we should be able to apply a like degree of American ingenuity and engineering skills to overcome the pollution problem.

There's no denying a certain degree of logic to these arguments. On the other hand, there is the very real question as to whether it will be possible, or indeed feasible, to mobilize the same degree of national effort to clean up the environment that it took to translate into reality President John F. Kennedy's vow to put man on the moon before the end of the '60's.

Nevertheless, efforts to reduce the emission of pollutants from various sources are rapidly gaining momentum. The results of such efforts may not be as dramatic as a space shot nor the results immediately evident, but at least a start has been made.

Fig. 6.6   Original passage

footnote it, even if it contains only a couple of words. To shorten a quote, you can leave out some of the original words; mark such an omission, or ellipsis, with three spaced periods if the omitted words are in the middle of a sentence, and four spaced periods if they are at the end of the sentence. You can also insert short explanations into quoted material, but you must enclose your own words in square brackets. These techniques are exemplified in Fig. 6.7, a quotation from the original passage shown in Fig. 6.6.

When paraphrasing, you retain the meaning of a passage but put it into your own words. You cannot merely substitute synonyms and alter the author's sentence structure; if you use the author's statements that extensively, you must quote them instead of paraphrase them. Though not placed in quotation marks, paraphrases are footnoted. To avoid any doubt about where your ideas leave off and the paraphrased ideas begin, identify the author at the beginning of the paraphrase, as shown in Fig. 6.8.

When quoting and paraphrasing, you have some obligations to both the original author and the report reader:

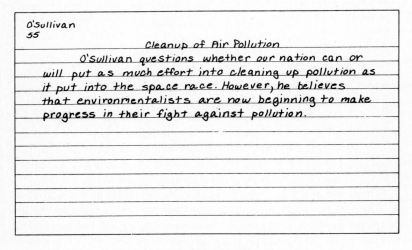

O'Sullivan
55

Cleanup of Air Pollution

"They [environmentalists] also frequently cite the fact that if we can land men on the moon and bring them safely back to earth again, we should be able to apply a like degree of American ingenuity and engineering skills to overcome the pollution problem.....On the other hand, there is the very real question as to whether it will be possible, or indeed feasible, to mobilize the same degree of national effort to clean up the environment that it took to ... put man on the moon before the end of the '60's."

Fig. 6.7   Quotation from passage

O'Sullivan
55

Cleanup of Air Pollution

O'Sullivan questions whether our nation can or will put as much effort into cleaning up pollution as it put into the space race. However, he believes that environmentalists are now beginning to make progress in their fight against pollution.

Fig. 6.8   Paraphrase of passage

1. Whenever in doubt about whether an idea is yours or another author's, give credit to the other author.
2. Do not quote or paraphrase in a way that misrepresents the original author's meaning. You must not distort the context of the author's words during the switch from a printed source to a report.
3. Never use extended paraphrases and rarely use extended quotes. If a report becomes nothing but other people's ideas thrown together, its objectives will not be met and the reader will have to conclude that you have done no independent thinking.

Documenting the Sources

The most widely used method of documentation follows the footnote-bibliography format of the MLA (Modern Language Association). This format is described here and exemplified in two student models at the end of the chapter.

The MLA method of documentation is one of many that are in use today; before writing a research report, find out the format preferred by your instructor. You may be referred to a format followed in a particular scholarly journal in your field or to a style sheet prepared by your department.

FOOTNOTES

When using the footnote-bibliography format, place consecutive footnote numbers immediately after each quote or paraphrase. Then place a corresponding number and a statement of the source at the bottom of the page. If you use a footnote to provide a definition or an explanation, number it just as you would a reference to a source. Following are sample footnotes for the most common sources of material; a complete list is available in the *MLA Handbook*, published in 1977.

Book:

[1]Barry Commoner, <u>Science and Survival</u> (New York: Viking Press, 1966), p. 147.

[If the report contains a bibliography page, the name of the publisher can be omitted: (New York, 1966), p. 147.]

Book by two authors:

²Trevor Lewis and Lionel R. Taylor, <u>Introduction</u> <u>to</u>
<u>Experimental</u> <u>Ecology</u> (New York: Academic Press, 1967), p.
230.

(The authors are named in the order they appear on the book's title page. If
the book had more than two authors, the footnote would begin: Trevor Lewis
and others, *Introduction* . . . .)

Second or subsequent edition of a book:

³Raymond F. Dasmann, <u>Environmental</u> <u>Conservation</u>, 2nd ed.
(New York: Wiley, 1968), p. 112.

Essay in a volume of essays:

⁴Georg Borgstrom, "The Harvest of Seas: How Fruitful and
for Whom?" in <u>The</u> <u>Environmental</u> <u>Crisis</u>: <u>Man's</u> <u>Struggle</u> <u>to</u>
<u>Live</u> <u>with</u> <u>Himself</u>, ed. Harold W. Helfrich, Jr. (New Haven:
Yale University Press, 1970), p. 82.

Reference to a work already cited:

⁵Commoner, p. 155.

(If two works by Commoner were being used, this footnote would also include
the title or a shortened title of the book: Commoner, *Science and Survival*,
p. 155.)

Article in a weekly magazine:

⁶Dermot A. O'Sullivan, "Air Pollution," <u>Chemical</u> <u>&</u>
<u>Engineering</u> <u>News</u>, 8 June 1970, p. 38.

Article in a monthly magazine:

⁷Gordon Barclay, "Solar Houses," <u>Architectural</u> <u>Record</u>,
November 1976, p. 115.

Article in an encyclopedia:

⁸"Animal Ecology," <u>Encyclopaedia</u> <u>Britannica</u>, 1965.

(If the article were signed, the author's name would precede the article. Volume and page numbers are not needed for single-page articles arranged alphabetically in reference books; however, volume and page numbers must be given if the citation is to only one page of a multipage article.)

Newspaper Article:

[9]Alvin Nagelberg, "Solar Energy Emerges from the Lab," Chicago Tribune, 1 June 1975, Sec. 2, p. 17, col. 1.

Pamphlet, Bulletin, or Report:

[10]L. N. Smith and W. L. Miller, Manpower Supply in Waste-Water Treatment Plants, Technical Report No. 15, Purdue University Water Resources Research Center (Lafayette, Indiana, 1970), p. 3.

Interview:

[11]Personal interview with Jane R. Shoup, Associate Professor of Biology, Purdue University—Calumet Campus, 15 March 1976.

(If the interview were by telephone, the footnote would begin, "Telephone interview with . . . .")

BIBLIOGRAPHY

On your bibliography page at the end of the report, list the sources of material in alphabetical order according to the last name of the author. If the author is not named, list the article according to the first letter of the title, not counting A, An, or The. A bibliography may contain all sources consulted or only those used, depending on the instructor. Each of the following bibliographical entries corresponds to one of the sample footnotes.

"Animal Ecology." Encyclopaedia Britannica, 1965.

Barclay, Gordon. "Solar Houses." Architectural Record, November 1976, pp. 111-118.

Borgstrom, Georg. "The Harvest of Seas: How Fruitful and for Whom?" The Environmental Crisis: Man's Struggle to Live with Himself. Ed. Harold W. Helfrich, Jr. New Haven: Yale University Press, 1970, pp. 65–84.

Commoner, Barry. Science and Survival. New York: Viking Press, 1966.

Dasmann, Raymond F. Environmental Conservation. 2nd ed. New York: Wiley, 1968.

Lewis, Trevor, and Lionel R. Taylor. Introduction to Experimental Ecology. New York: Academic Press, 1967.

Nagelberg, Alvin. "Solar Energy Emerges from the Lab." Chicago Tribune, 1 June 1975. Sec. 2, p. 17, cols. 1–4.

O'Sullivan, Dermot A. "Air Pollution." Chemical & Engineering News, 8 June 1970, pp. 38–41.

Shoup, Jane R. Personal interview. 15 March 1976.

Smith, L. N., and W. L. Miller. Manpower Supply in Waste-Water Treatment Plants. Technical Report No. 15, Purdue University Water Resources Research Center. Lafayette, Indiana, 1970.

Typing the Research Report

When you use a quotation that requires more than four typed lines, indent and single-space it as shown in the two models at the end of this chapter. You need not place quotation marks around such quotations because the physical layout tells the reader that the information is quoted; however, you must place a footnote number at the end of the excerpt. This method of indenting applies only to quotations; it cannot be used for paraphrases.

Endnotes are sometimes used instead of footnotes. Endnotes are identical to footnotes, but they are placed on a separate page at the end of the report instead of at the foot of each page. The first model at the end of this chapter contains footnotes; the second has endnotes.

Models

COMPUTERS THAT LISTEN

INTRODUCTION

Until recently, computers have required their operators to use special "languages" and coding schemes in order to enter information into the computer system for processing. The only communication media between human and machine were punched cards, punched paper tape, magnetic devices such as tapes or disks, and direct electrical impulses such as those on process-control computers.

At Bell Research Laboratories, a new method of communication with computers has been developed:

> With the widespread growth in the use of digital computers, there has been an increasing need for man to be able to communicate with machines in a manner more naturally suited to humans. The realization of this need has motivated a great deal of research in automatic recognition of speech by computer.[1]

Automatic speech recognition allows the human voice to be used directly by the computer as input information. The new developments in automatic speech-recognition systems are gradually finding applications in the education and business worlds.

[1]M. R. Sambur and L. R. Rabiner, "Speaker Independent Digit-Recognition System," Bell System Technical Journal, January 1975, p. 81.

SYSTEMS NOW ON THE MARKET

The two major suppliers of speech-recognition systems in the United States are Scope Electronics and Threshold Technology. Both systems must be "trained" by each of their users to understand the words in the user's command vocabulary, usually 30 to 35 words. This training period is necessary for the computer to learn each user's unique speech patterns. Although the systems can recognize only the speech patterns of users for which they are trained, variations in an individual's speed or loudness while speaking a certain word do not affect the systems' reliability. The systems correctly recognize words about 98% of the time, and even this error range can be eliminated with the use of display panels by the operator to insure that the system understands correctly.

Although the two systems are operated almost identically, the technology that each uses for identifying spoken words and translating them into signals for the computer is different. Human speech is an analog signal (a continuous progression of frequencies) but computers use digital signals (distinct on-and-off pulses), so the first step of both systems is to change the voice signal from analog to digital. In the Scope system, the voice signals are analyzed by noting

the relationship between frequency (or pitch) and time, and
then comparing this data with the stored data of the users'
recorded voices. In contrast, the Threshold system analyzes
the voice signal by looking for acoustic features such as
vowels, consonants, and abrupt bursts of energy. These tiny
features, called phonemes, are compared with the stored features of the users' voices.

APPLICATIONS

Current applications of voice-command systems, as
speech-recognition systems are often called, are concentrated
in situations where users cannot readily use their hands to
enter data in the conventional manner. This includes package
handling in airline terminals, warehouses, and post offices.
By freeing hands, the voice-command system sometimes allows
one person to do the work of two, and often with fewer errors
than keypunching causes.

Product Engineering reports that voice-command systems
are now being successfully used in quality-control applications, also. The system displays on a screen a series of
questions that must be answered by the inspector to assure a
certain quality. The inspector has his or her hands free to
make measurements that are directly transmitted to the com-

puter (by voice), thus eliminating several steps that might have led to errors.[2]

According to L. H. Rosenthal of Bell Labs, automatic speech recognition systems may someday be used extensively for security purposes. A person's voice may be all the identification needed to make a bank withdrawal, charge a purchase to a credit account at a department store, or gain access to restricted data files on a computer.[3]

CURRENT RESEARCH

One of the biggest limitations of current voice-command systems is the need to train them to recognize each user's voice. Researchers today are trying to program the computer to understand <u>any</u> voice, using a limited vocabulary. Scientists at Bell Labs have developed a system to accomplish this objective, using zero and the nine digits as its limited vocabulary:

> Two major problems had to be overcome before the computer could "understand" spoken numbers. First, speech quality differs from person to person, and the computer had to be programmed to recognize the same words as spoken by different individuals. Second, because human speech is continuous—the

[2]"Machines That Understand Spoken Commands Save Time and Money," <u>Product Engineering</u>, December 1975, p. 11.

[3]L. H. Rosenthal and others, "Automatic Voice Response; Interfacing Man with Machine," <u>IEEE Spectrum</u>, July 1974, p. 65.

words run together without pauses—the computer had
to be able to isolate individual words for identification.[4]

The Bell Labs team solved these problems by "self-nor-

malizing" the acoustical characteristics of the numerals;

that is, the characteristics of each individual's voice were

identified and taken into account by the computer program be-

fore the words were analyzed. This process resulted in cor-

rect results 98% of the time when the numerals were spoken in

a quiet room. Eventually, if sufficient improvements are

made in this technique, human voices instead of fingers may

direct telephone calls.

CONCLUSION

The recent developments in automatic speech—recognition

systems are just now finding their way into limited business

applications. Within the next decade, however, it is pre-

dicted that these "listening" computers will find widespread

uses in both industry and the home.

[4]"Progress Made in Programming Computers to Listen,"
Bell Laboratories Record, November 1975, p. 53.

BIBLIOGRAPHY

"Machines That Understand Spoken Commands Save Time and
Money." Product Engineering, December 1975, pp. 10-12.

This article describes two major voice-recognition systems
on sale in the United States today. It also describes major applications of the new equipment.

"Progress Made in Programming Computers to Listen." Bell
Laboratories Record, November 1975, p. 53.

This article describes the progress being made on various
projects on voice-recognition systems. It pays special
attention to the speaker-independent system developed at
Bell Labs.

Rosenthal, L. H., and others. "Automatic Voice Response; Interfacing Man and Machine." IEEE Spectrum. July 1974,
pp. 61-68.

This article focuses on voice-response systems and their
possible uses in business. Also included is a section on
how these systems can be used in conjunction with speaker-
recognition systems to provide a computer that both lis-
tens and talks.

Sambur, M. R., and L. R. Rabiner. "Speaker Independent
Digit-Recognition System." Bell System Technical Journal,
January 1975, pp. 81-102.

Sambur and Rabiner thoroughly describe their speaker-independent digit-recognition system. The authors describe in
great detail the acoustical theories behind their research.

SOLAR ENERGY FOR HOMES

INTRODUCTION

As monthly utility bills have risen sharply, the search for an economical method of heating and cooling homes has brought about new ideas and resurrected some old ones. One of those old ideas is the use of solar energy. This report will discuss basic theory and current research into solar energy for residential use.

BASIC THEORY

The theory behind solar heating and air conditioning is very simple. Flat-plate collectors on the roof of the house use mirror-surfaced sheet metal to reflect sunlight onto water pipes. The collector plates are heavily insulated and covered with sturdy, clear-glass facing to minimize heat loss, prevent damage, and provide ease in cleaning. Solar radiation heats water circulating through the pipes to near the boiling point: the hot water is stored under pressure in an insulated tank.[1]

When a room requires heat, hot water from the storage tank is pumped through conventional heating pipes, and the heat is radiated into the air. When cooling is necessary, the hot water operates as an air refrigeration system. Both

heating and cooling are controlled by conventional thermo-
stats. The insulated storage tank provides a reservoir of
hot water for use at night and on cloudy days when solar energy does not reach the collector plates.

A home must be solar oriented, or positioned, to maxi-
mize the amount of heat gained from solar radiation. In the
U.S., collectors are mounted to face south at a slope perpendicular to the sun's rays. At present, the area of the solar
collectors must be 50% of the floorspace area to be heated
and cooled.[2]

CURRENT RESEARCH

In cooperation with Honeywell, ERDA (U.S. Energy Research and Development Administration) developed a transport-
able solar laboratory. ERDA has tested the lab in various
climates throughout the nation to discover the practical potential of solar energy for heating and cooling homes. Although some climates have been shown to be less acceptable
for solar power than others, the experiments taken in the lab
have proved the feasibility of solar homes.[3]

The effectiveness of solar homes in cold climates has
also been demonstrated by Gordon Barclay, an architect who
built a solar home in Vermont:

> Okemo Mountain, in Ludlow, Vermont, has an eleva-
> tion of 3,700 ft, winter temperatures that some-

> times do not rise above 15 F for periods as long as
> 21 days, and a winter sunshine factor in the range
> of 40 to 45 per cent. . . . The home temperature
> during the coldest months was maintained at approx-
> imately 69 to 72 degrees F. . . . When springtime
> arrived the home had received approximately 95 per
> cent of its heat from the sun.[4]

Successful solar heating under such adverse conditions shows

the usefulness of solar homes in the U.S.

CONCLUSION

Engineers are now faced with the problem of reducing

the cost of solar components to make solar homes more econom-

ical to build. Although estimates vary, most experts say

that the price of solar collectors and heat-storage facili-

ties for a three-bedroom house presently exceeds $5,000.

As components are refined and become mass produced,

however, solar heating and cooling systems should become less

costly. By 1985, according to ERDA's projections, solar en-

ergy will provide 0.8% of the nation's total energy needs.

This figure will rise to 7% in the year 2000 and reach 25% by

the year 2020.[5]

ENDNOTES

[1]"Various Uses for Sun Power," Chicago Tribune, 9 August 1976, Sec. 1B, p. 17, cols. 4–5.

[2]"Solar Orientation," Dictionary of Architecture and Construction, 1975.

[3]"Various Uses for Sun Power," col. 4.

[4]Gordon Barclay, "Solar Houses," Architectural Records, Nov. 1976, p. 115.

[5]"Solar Heating Is Here and Booming," U.S. News & World Report, 29 March 1976, p. 32.

BIBLIOGRAPHY

Barclay, Gordon. "Solar Houses." <u>Architectural</u> <u>Record</u>, Nov.
    1976, pp. 111–118.

    Barclay describes a solar home in an area of Vermont
    where temperatures do not rise above 15 degrees for pe-
    riods as long as 21 days. The solar heating system sup-
    plied 95% of the heat required from September through
    April.

"Solar Heating is Here and Booming." <u>U.S.</u> <u>News</u> <u>&</u> <u>World</u>
    <u>Report</u>, 29 March 1976, pp. 30–32.

    Solar energy's future is evaluated by the U.S. Energy Re-
    search and Development Administration in this article.
    Cost estimates are also provided by the Solar Energy In-
    dustries Association.

"Solar Orientation." <u>Dictionary</u> <u>of</u> <u>Architecture</u> <u>and</u>
    <u>Construction</u>, 1975.

    The dictionary provides definitions of many solar energy
    terms. One such definition is of solar orientation, the
    placement of a building so that the sun's radiation is
    maximized or minimized to effectively operate a solar en-
    ergy system.

"Various Uses for Sun Power." <u>Chicago</u> <u>Tribune</u>, 9 August
    1976, Sec. 1B, p. 17, cols. 4–5.

    This article describes tests conducted in a portable so-
    lar laboratory developed by the federal government. The
    lab was used to test climates throughout the nation to de-
    termine if those climates were suitable for solar power.

Writing Assignment

Write a 1000-word research report explaining a recent innovation or development in your field to an uninformed reader. The report should contain quotes and paraphrases from journals in your field. Avoid writing a report that contains nothing but quotes and paraphrases by selecting a topic about which you have some background knowledge. For example, perhaps your report can be about a new mechanism that might eventually make one of the many mechanisms in your field obsolete.

Your research report should have an introduction that gives background information and a conclusion that indicates the overall significance of the development. The information you present in the body of the report will depend on your topic. Some kinds of information that you might present include (1) problems and potential solutions regarding the development; (2) the development's applications to your field, to industry, or to society in general; and (3) the development's relationship to general trends in your field or in industry.

Your readers will have no rigid preconceptions about the information or headings contained in the body of your report, but they will expect the report to be logically organized, clear, and properly documented.

Exercise

Examine *Ulrich's International Periodicals Directory* or *The Standard Periodical Directory* in the library. Each lists the periodicals published for all fields. Select three of the most popular periodicals in your technical area and inspect several copies of each. Then write a one-paragraph evaluation of each periodical's potential usefulness to you following graduation.

7 Illustrating

Visual media have had a tremendous impact on everyone's communication skills. Mainly because of television, people grasp and interpret visual information more readily than printed information. Illustrations not only speed the communication process, which is important in today's fast-paced industrial society, but serve as common denominators for report readers. Visual support accelerates an informed reader's rate of comprehension and helps insure the understanding of uninformed readers.

Another contribution of illustrations is the visual attractiveness they lend to reports. There is still a bit of the comic book reader in most people, and anything that contains illustrations simply looks understandable. For example, when you flip through a textbook for the first time, you usually make a judgment. If the book contains nothing but page after page of paragraphs, you suspect that it will be difficult to understand. If the book contains illustrations, you get the opposite impression, and that impression is accurate if the visuals actually clarify the text. Keep in mind, however, that illustrations selected just for their attractiveness are like words chosen for their impressiveness. Words and visuals have only one purpose: to communicate information.

As a report writer, you select illustrative material by applying the same principles you use for choosing words. Instead of inserting the first visual that comes to mind, you consider the report's subject and reader. Then, taking advantage of the options discussed in this chapter, you decide which type of illustration will be successful.

This chapter discusses five kinds of illustrations: photographs, drawings, diagrams, tables, and graphs. The first part of the chapter covers photographs, drawings, and diagrams; most of these figures are exemplified by material from the Heath Company's *Assembly Manual for the Citizen's Band Transceiver, Model GW-22*. Heath products are sold in kits, called Heathkits, for assembly by people with various levels of technical ability. Therefore, the manuals accompanying Heathkits must contain figures that effectively reinforce complex technical information.

Photographs

A photograph's advantage, the ability to show details clearly, can also be its disadvantage. At some points in a report, too much detail can confuse the reader by detracting from what you want to emphasize. However, photography departments in industry are becoming very adept at arranging the subjects of photographs or airbrushing extraneous material out of them. Unless prepared skillfully, a photo is not inherently better than a drawing.

Drawings

As was indicated previously, you have control over the content of your drawings and can therefore insure that they are integrated with the text of your report. Drawings are particularly helpful for readers who must become familiar with the physical characteristics of a mechanism before being able to understand its operation.

Having considered the reader in choosing an overall type of visual—drawings—you consider the object when deciding whether a cutaway, exploded, or cross-sectional drawing will work best.

CUTAWAY DRAWINGS

A cutaway view removes a portion of the mechanism's casing, showing the inside of the mechanism, revealing the relationships between the inner parts, and clarifying the position of the interior assembly in relation to its housing. Figure 7.1 exemplifies this, although a cutaway is more often used for smaller mechanisms.

EXPLODED DRAWINGS

As the word implies, an exploded view blows an object apart but maintains the arrangement of its parts, as in Fig. 7.2. Exploded drawings are useful when

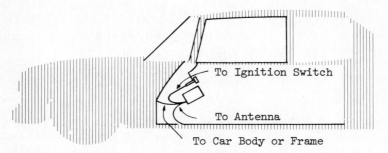

To Ignition Switch

To Antenna

To Car Body or Frame

Fig. 7.1 Cutaway drawing

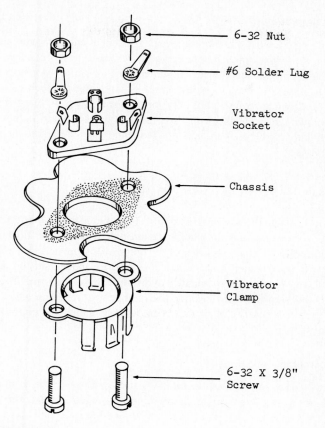

6-32 Nut

#6 Solder Lug

Vibrator
Socket

Chassis

Vibrator
Clamp

6-32 X 3/8"
Screw

Fig. 7.2   Exploded drawing

you want to show the internal parts of a small and intricate object or explain
how it is assembled.

CROSS-SECTIONAL DRAWINGS

Cross-sectional views are similar to cutaway views except that they cut the
entire assembled object, both exterior and interior, in half. In technical terms,
the object is cut at right angles to its axis. A cross-sectional view shows the
size and relationship of all the parts. Two views of the same object, front and
side views, for example, are often placed beside each other to give the reader
an additional perspective of the object.

PROCESS DRAWINGS

Another type of visual that emphasizes physical characteristics can arbitrarily be called a process drawing. It is generally reserved for manuals, where the directions sometimes need visual reinforcement. A step-by-step sequence of drawings assists the reader's performance of an operation, such as soldering (Fig. 7.3). Operations involving only mechanisms, such as the strokes of a piston drawn on page 58 of this text, as well as processes including both workers and machines, can be depicted either in simple drawings or in photographs.

Diagrams

Diagrams are actually a kind of drawing. The significant difference between the two is that diagrams communicate through symbols and do not try to show the physical characteristics of an object. This presents an immediate problem: Is your reader well enough informed to understand the technical symbols? To achieve communication, you must make the symbols conform to the reader's knowledge. This becomes a matter of selecting either a block diagram or a schematic diagram, because both show the operation of a mechanism.

BLOCK DIAGRAMS

A football coach drawing a play on the blackboard uses symbols to indicate each player's assignment. In much the same way, a block diagram represents

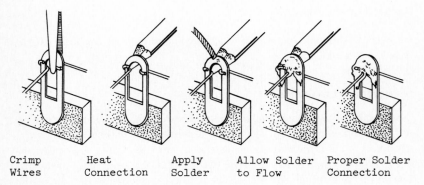

```
Crimp        Heat          Apply      Allow Solder    Proper Solder
Wires        Connection    Solder     to Flow         Connection
```

Fig. 7.3   Process drawing

Fig. 7.4   Block diagram

several sequences or stages in a mechanism's operation. Each block has a label identifying the function of the sequence, and arrows indicate transition between the sequences. Make sure that the labels on a diagram are meaningful to your reader and that the diagram and paragraphs of description are well integrated. The circuit diagram in Fig. 7.4, which represents the power supply sequences of the Heath Company's citizen's band transceiver, helps clarify the manual's description of the operation.

Block diagrams can also show the sequences of complex operations involving workers and machines. Commonly called a flow diagram or flow chart, this kind of diagram symbolically traces such processes as a factory's conversion of raw material into a finished product.

SCHEMATIC DIAGRAMS

The power-supply process shown in Fig. 7.4 is also represented in Fig. 7.5, a schematic diagram. These two diagrams appear almost side by side in the Heathkit manual to serve all readers. The schematic symbols are aimed at an informed reader, but they are clearly defined in the manual.

Schematics are used to represent systems in many fields, including electronics and hydraulics. Like blueprints, they are excellent time-saving devices if the reader understands them.

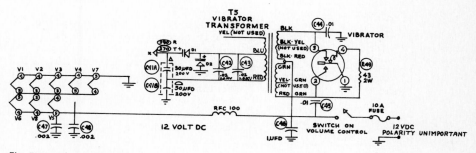

Fig. 7.5   Schematic diagram

Tables

Tables, which classify data in columns and rows, are used extensively in reports to present statistical information. Pages 81–82 of Chapter 5 demonstrated their effectiveness by contrasting data presented in a table with the same data presented in paragraph form.

Statistics are not the only type of information that can be tabulated. You can frequently categorize and tabulate written information to speed the communication process. The form of Fig. 7.6, a troubleshooting chart from a Heath Company manual, permits the information to be understood much faster than it would be if it were presented in paragraph form. When writing reports and manuals, try to recognize information that can be presented more simply and clearly in tabular form.

The elements and organization of a table are shown by Fig. 7.7 and Fig. 7.8, the first identifying the elements and the second exemplifying them. The eight elements are the following:

1. Table number: Tables are numbered consecutively throughout a report. Of all the illustrations discussed in this chapter, tables are the only ones whose number and title appear *above* them.

DIFFICULTY	POSSIBLE CAUSE
Receiver section dead.	Check V1, V2, V3, V4, V7, and V8. Wiring error. Faulty speaker. Faulty receiver crystal. Crystal oscillator coil mistuned.
Receiver section weak.	Check V1, V2, and V3. Antenna, RF or IF coils mistuned. Faulty antenna or connecting cable.
Transmitter appears dead.	Check V5 and V6. Wiring error. Recheck oscillator, driver, and final tank coil tuning. Dummy load shorted or open.

Fig. 7.6 Troubleshooting chart

(1)   TABLE NUMBER   (2)   TITLE

(4) Stub Heading	(3) Column Heading	Column Heading
(5) Line Heading		
(6) Subheading	Data	Data
Subheading	Data	Data
Subheading	Data	Data
Line Heading	Data	Data
Line Heading	Data	Data

(7)   Explanatory note.

(8)   Source.

Fig. 7.7   Elements of a table

(1)   TABLE 1:   (2)   COST OF SPEED DRIVES

(4) Initial Costs	(3) AC Drive	DC Drive
(5) Purchase		
(6) Speed Drive[a]	$ 76,000	$110,000
Air Damper	1,500	—
Tachometer	2,000	—
Installation	23,000	12,000
Total	$102,500	$122,000

(7)   [a]One-year warranties are provided with both drives.

(8)   Source: John K. Smith, "Speed Drives," Production Journal, July 1977, p. 52.

Fig. 7.8   Sample table

2. Title: The title of a table should concisely identify the data contained in the table.
3. Column Headings: Column headings classify the data below them.
4. Stub Heading: A stub heading classifies the line headings that appear below it.
5. Line Headings: A line heading such as "Purchase" classifies the subheadings below it. In the absence of subheadings, a line heading such as "Installation" or "Total" identifies the contents of a horizontal row of data.
6. Subheadings: Subheadings identify the contents of horizontal rows of data.
7. Explanatory Notes: Notes may be used to clarify portions of a table. They are indicated by raised, lower-case letters within the table and at the beginning of the note.
8. Source: When a table is from a source outside your company, you should name the source. The form of a source statement is identical to the footnote form described in Chapter 6.

Graphs

Statistical data that can be placed in a table can also be indicated by a graph. Bar graphs, circle graphs (pie charts), and line graphs (curves) present information more dramatically than tables; however, they often do not show specific numbers as tables do. In other words, graphs are more visual than tables but often less precise. Therefore, you should determine whether a table or graph is more appropriate to your reader's needs. If you decide to present the information graphically, you must determine if a bar, circle, or line graph is best.

BAR GRAPHS

Bar graphs are effective for showing comparative quantities. Figure 7.9 is a graph from IBM Corporation's *Annual Report 1976* showing capital expenditures over a 5-year period. The years are shown beneath each bar, and the vertical scale indicates capital expenditures, with each grid representing $350 million. The height of the bars shows at a glance the relative increase or decrease during the 5 years. The graph could be made both visual *and* precise if dollar figures were placed immediately above each bar.

Bar graphs can be used to show relative quantities, such as the number of employees, products manufactured, accidents, or hours spent repairing equip-

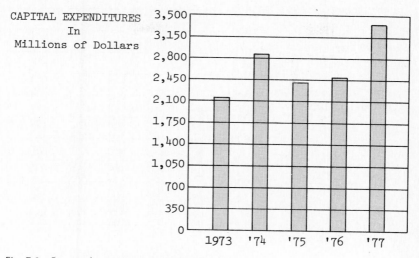

CAPITAL EXPENDITURES
In
Millions of Dollars

Fig. 7.9 Bar graph. Courtesy IBM Corporation.

ment over a period of time. Vertical bars are used except to compare such things as distance, for which horizontal bars are more appropriate.

Segmented bar graphs are used to compare not only total amounts over a period of time but also components of total amounts. Figure 7.10 segments IBM Corporation's gross income for each of 5 years into income gained from sales and from rentals and services during each year. Again, the strength of such a graph is its visual quality, and the weakness is its lack of precise figures.

Figure 7.11 shows how more than one bar can be used to compare quantities over a certain period. The first bar for each year shows IBM's net earnings, and the second bar shows the amount of cash dividends paid to stockholders. Such a graph is effective for comparing such items as the number of male versus female employees or the relative sales of two (or more) products over a period of time.

CIRCLE GRAPHS

Circle graphs are effective for simultaneously comparing the components of a whole to each other and to the whole. In Fig. 7.12, from *American Airlines' Annual Report 1975*, each circle represents a dollar. The circles are divided

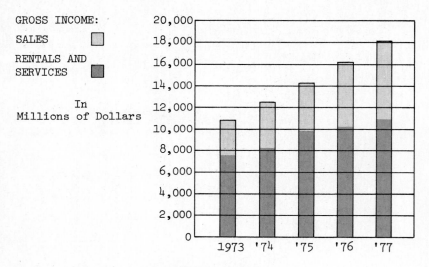

Fig. 7.10   Segmented bar graph. Courtesy IBM Corporation.

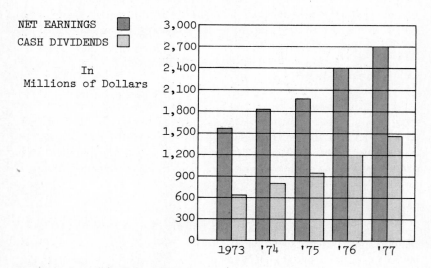

Fig. 7.11   Bar graph. Courtesy IBM Corporation.

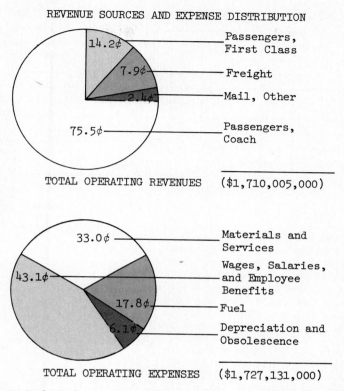

REVENUE SOURCES AND EXPENSE DISTRIBUTION

14.2¢ — Passengers, First Class

7.9¢ — Freight

2.4¢ — Mail, Other

75.5¢ — Passengers, Coach

TOTAL OPERATING REVENUES   ($1,710,005,000)

33.0¢ — Materials and Services

43.1¢ — Wages, Salaries, and Employee Benefits

17.8¢ — Fuel

6.1¢ — Depreciation and Obsolescence

TOTAL OPERATING EXPENSES   ($1,727,131,000)

Fig. 7.12 Circle graphs. American Airlines Annual Report to Stockholders 1975.

into slices showing how many cents of each revenue dollar were gained from certain sources, and how many cents of each expense dollar were spent in certain ways. Good circle graphs are clearly labeled and show the unit values of each slice.

LINE GRAPHS

A line graph, or curve, allows data to be plotted in order to show a trend. Figure 7.13, from the *1974 Annual Report* of Eastern Air Lines, Inc., is typical because the independent variable (time) is plotted horizontally, along the abscissa. The dependent variable (cost of gas) is plotted vertically, along the

Cost per Gallon in Cents --12 Months Moving Average

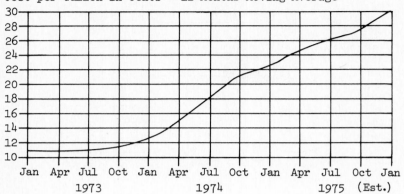

RISING FUEL COSTS
By year-end 1974, the average
price per gallon of fuel paid by
Eastern was more than double
what it was at year-end 1972. In
1975, fuel prices are expected to
continue their upward climb so
that by the end of the year they will
be almost triple what they were at
year-end 1972. Even though
Eastern will use an estimated 6.5%
less fuel in 1975 than in 1972,
it will cost an estimated
$173 million more.

Fig. 7.13   Line graph. Courtesy Eastern Airlines.

ordinate. The graph shows how the cost of gas (dependent) has been affected over a period of time (independent). A line graph's ability to show the relationship between variables makes it a most useful kind of technical illustration.

More than one line can be plotted on a line graph for the purpose of comparing. Figure 7.14, from the *1976 Annual Report, Southern California Edison Company*, compares the company's original (1973) and revised (1977) forecasts of future demands for electricity.

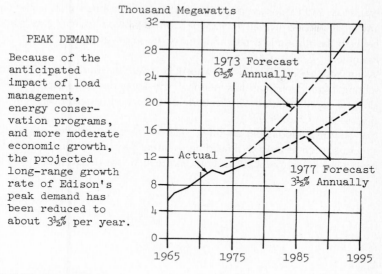

PEAK DEMAND

Because of the anticipated impact of load management, energy conservation programs, and more moderate economic growth, the projected long-range growth rate of Edison's peak demand has been reduced to about $3\frac{1}{2}\%$ per year.

Fig. 7.14   Line graph

Guidelines for Effective Illustrations

1. Identify all illustrations as either tables or figures; anything not a table is a figure, no matter what form it takes. Place numbers and titles directly above tables and directly below figures.
2. Integrate illustrations into the report at logical and convenient places. An illustration should not precede your discussion of it; if it does, the reader might spend 15 minutes trying to understand the illustration, only to discover that you have clarified it on the next page.
3. Refer to each illustration in the text of the report, even if the illustration is right beside or below your discussion of it. Also, explain or interpret each illustration instead of letting it speak for itself.
4. Do not clutter an illustration with too many words, causing it to lose its visual quality and impact. When labels are needed for numerous parts in a figure, letter or number the parts, place the labels in a key next to the drawing, and use the letters or numbers for textual references to the parts, thereby saving space in both the illustration and the text.
5. Do not assume that an illustration appearing in a technical journal will automatically be effective in your report. As a report writer, your purpose,

emphasis, and reader are often quite different from those of the technical article. If you do use an illustration from a printed source, give credit beneath the illustration, providing the same information you would in a footnote.

6. When your report contains an appendix, exercise your options by presenting highly technical illustrations in the appendix and simplified versions of the illustrations in the body of the report. (Appendixes will be discussed in Chapter 9).

7. Not only illustrations but anything different from the ordinary paragraph structure, such as a list or a heading, is visual because it calls attention to itself. Give visual emphasis only to information that needs or deserves it.

Section Three
Report Formats

The techniques described in Section Two apply to both informal and formal reports. The formats of these reports are quite different, however. Section Three separates informal and formal reports, explains why they must be approached differently, and emphasizes methods of organizing them to communicate with their respective readers.

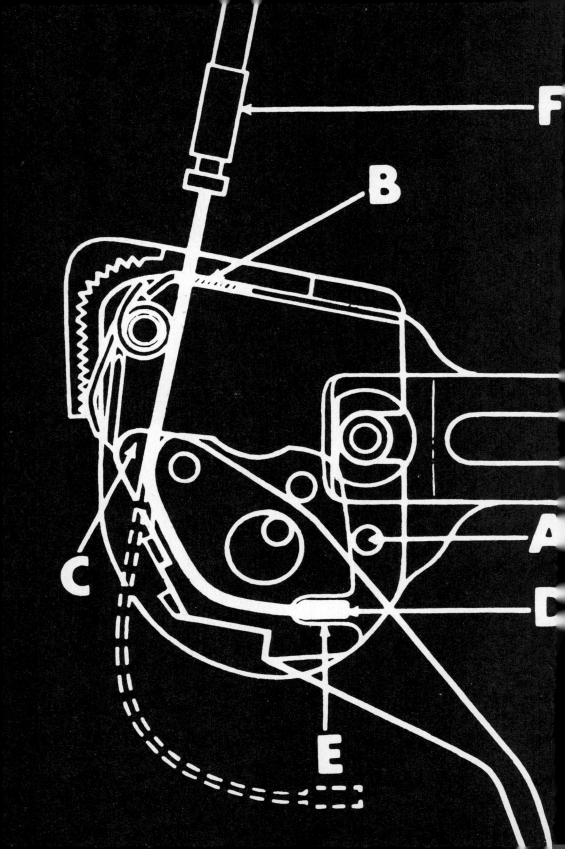

8 Informal Reports and Memorandums

THE day-to-day operation of a company depends on informal reports that circulate within and among its departments. These reports carry the results of investigations and convey information about products, methods, and equipment. The most obvious difference between informal and formal reports is their format. Formal reports have title pages, tables of contents, and abstracts. Informal reports, whose formats will be emphasized in this chapter, generally have none of these; in fact, most informal reports are written on printed memorandum forms. The differences in the content of these two kinds of reports are the following:

Table 8.1   Differences between informal and formal reports

	Informal reports	Formal reports
Scope	attack either a problem of narrow scope or a limited number of aspects of a large problem	attack all aspects of a large problem
Purpose	to give departmental information or action, but occasionally to relay preliminary or partial results to management	to play a role in a decision-making process at management level
Readers	are often aimed at only one person, who is usually technically informed	are read by several people, from as many management areas as are involved with the problem, some of whom fall into the technically uninformed category
Emphasis	emphasize the results of the investigation and the procedure by which data and results were obtained	emphasize the results of the investigation; the readers are interested in procedure only to the extent of knowing that the results are valid

The most significant of these differences is the readers' levels of technical sophistication. Invariably, some of a formal report's readers possess limited technical knowledge; therefore, you place highly technical information in the appendix of the report. On the other hand, an informal report's reader generally has a strong technical background. He or she is often someone who has worked up to a supervisory position in a department.

Nevertheless, an informal-report writer cannot assume that the reader has total familiarity with the subject of the report. Every report involves investigation, thought, and writing, and the reader does not accompany the writer during all these stages. If you have done your job well, you are inevitably more informed about the subject than your reader; keeping this in mind, you must present your information as clearly as possible.

Memorandums

Most firms use printed memorandum forms for brief, handwritten notes whose verbal counterparts are messages conveyed by telephone. These notes, which conform to the strict definition of a memorandum as a written reminder, cannot properly be called reports. Memorandum forms are so convenient, however, that most companies use them to communicate the results of investigations. In other words, you will use memorandum forms to convey information that falls into the informal-reports category.

A typical memorandum form used by Jonaitis Engineering Company is shown in Fig. 8.1; notice that it cautions writers to limit the scope of their information. The top of a memorandum form provides space for your name, the name of your reader, your subject, and the date. Depending on company practice, you may place the title and/or department in the space provided. The subject of your memorandum, which is usually typed in capital letters, should be as specific as possible. Within the body of lengthy memorandums, use headings to emphasize major sections, and reinforce your information with tables or figures. Place your signature or initials above your typed name at the end of the memo.

The importance of writing effective memorandums cannot be overemphasized. Because memos communicate the information necessary to keep a company running smoothly, they must be written clearly. Often they must be written quickly; do not be surprised if your supervisor says, "Send me a memo on that this afternoon." Memos test your ability to analyze a problem quickly and to write a concise, accurate solution. Therefore, your ability to handle memorandums tells your reader a great deal about your potential as a problem-solver and decision-maker.

The remainder of this chapter discusses and exemplifies informal recommendation reports and laboratory reports, two of the most important informal reports. These reports take the form of memorandums, as they frequently do in industry.

JEC JONAITIS ENGINEERING COMPANY
1715 Mandel Road, Chicago, Illinois 60646

COVER ONLY ONE SUBJECT IN EACH MEMORANDUM

TO: ______________________ DATE: ______________

____

FROM: ______________________

____

SUBJECT: ______________________________________

Fig. 8.1 Memorandum form

Informal Recommendation Reports

The technique described in Chapter 5 for interpreting statistics applies to informal recommendation reports and formal feasibility reports (Chapter 11), which both require the evaluation of alternatives. In an informal recommendation report, your objective might be to select the better of two methods, or to determine which of three types of equipment your company should purchase. You begin by gathering information about the alternatives from printed material, observation, or interviews. Then you select criteria by which to judge the alternatives. The most important criterion is generally cost, but sometimes you must consider other factors, such as efficiency or capability. You should use the following format for informal recommendation reports:

Introduction
  Purpose
  Problem
  Scope
Body
  Presentation and Interpretation of Data
Conclusions
Recommendations

Throughout the report, you use an inductive approach. You take the reader through statements of the report's purpose, problem, and scope in the introduction. Then, after devoting the body of the report to the presentation and discussion of data, you state your conclusions and recommendations.

This approach is exemplified in the following student report. The writer determines whether a proposed system would be more economical than the one currently used by the company. In the body of the report, he first demonstrates that the proposed system would reduce expenses. He then discusses the costs of installing the new system. Finally, he recommends purchase of the proposed system because it will quickly pay for itself.

MEMORANDUM

KAUZLARICH CONSTRUCTION COMPANY

TO: Mr. Dan Koeritz   DATE: September 19, 1978
        Manager
FROM: Jim Biggs
        Production Department

SUBJECT: RECOMMENDATION OF A METHOD FOR PURCHASING PAINT

INTRODUCTION

PURPOSE

The purpose of this report is to determine whether a
proposed method for purchasing paint would reduce costs.

PROBLEM

Because all steel strapping is painted to prevent corro-
sion, paint represents a major expense for the company. At
this time, the company purchases all its paint in 55-gallon
drums. It has been proposed that the paint be purchased by
tank truck at a saving of 10 cents per gallon.

The proposal would require installation of two permanent
650-gallon storage tanks at each of the company's two high-
volume machines. Also, twenty-eight 450-gallon portable
storage tanks would be required to maintain an adequate paint
inventory.

-1-

SCOPE

The only factor considered in this report is cost. The
report first compares annual paint costs under the present
and proposed systems. Then, expenses for purchasing and installing the proposed system are covered.

ANNUAL PAINT COSTS

The firm currently spends $2.18 per gallon of paint. Al-
though the price of paint fluctuates, a 10-cent-per-gallon
saving will always be given for tank-truck purchases. The
two high-volume machines use 150,000 gallons annually. As
Table 1 shows, this results in an annual cost of $327,000 un-
der the present system as opposed to $312,000 using the pro-
posed system. Thus, the proposed system would save $15,000
per year.

TABLE 1: ANNUAL PAINT COSTS

	Present	Proposed
Annual usage	150,000	150,000
Cost per gallon	$ 2.18	$ 2.08
Annual cost	$327,000	$312,000
Annual saving		$ 15,000

COST OF EQUIPMENT AND INSTALLATION

Costs for equipment and installation will be considered
only for the proposed system because these costs do not apply

-2-

to the present system. The proposed system requires that a
permanent tank be installed at each of the company's two
high—volume machines. Also, 28 portable tanks are needed to
maintain a 1—month inventory. The prices for the tanks
(Table 2) are $800 per permanent tank and $600 per portable
tank, and each of the permanent tanks would cost $1300 to in-
stall. The total cost for the proposed system would be
$21,000.

TABLE 2: COST OF INSTALLATION AND EQUIPMENT

Permanent tanks	Units	Total
Cost	2	$ 1,600
Installation	2	$ 2,600
Temporary tanks	28	$16,800
Total cost		$21,000

CONCLUSIONS

The proposed method of purchasing by tank truck would
cut $15,000 from yearly paint costs. The cost of implementing
this new system would be $21,000. Thus, the proposed system
would operate at a $6000 loss the first year, but it would
save $15,000 every year thereafter.

RECOMMENDATION

Because a $15,000 saving can be made annually after the
first year, adoption of the proposed tank–truck system is
recommended.

Jim Biggs

Jim Biggs

—4—

Laboratory Reports

The function of a laboratory report is to communicate information gained
through laboratory tests, which are the most rigid of all the data-gathering
methods. The accuracy of the test results depends on the procedure used
during testing; therefore, the laboratory report places special emphasis on
apparatus and procedure, as the following format indicates:

Introduction
  Purpose
  Problem
  Scope
  Apparatus (or Equipment)
  Procedure
Body
  Presentation and Interpretation of Data (often called Discussion)
Conclusions
Recommendations

A laboratory report's descriptions of apparatus and procedure affirm the accuracy of the data in the body of the report. Not all laboratory reports include
recommendations, but the following student report attempts to determine
whether a proposed concrete mix is strong enough to be used in a particular
construction project. In technical terms, the specifications call for concrete
that can withstand 5000 psi (pounds per square inch); this specification, or
criterion, receives a major heading in the body of the report. The writer does
not include a scope statement because his statement of the problem clarifies
the report's narrow scope.

MEMORANDUM

GARLITZ CEMENT COMPANY

TO: Ms. Ann Randall   DATE: August 5, 1978
        Manager, Testing Lab
FROM: Clinton Hare
        Senior Engineer

SUBJECT: TESTS ON CONCRETE FOR CANYON PROJECT

INTRODUCTION

PURPOSE

The purpose of this report is to present results of
tests on the concrete mix proposed for the Canyon Project and
to determine whether the mix is feasible.

PROBLEM

Specifications for the project state that the concrete,
which will be exposed to frequent freezing and thawing, must
be capable of withstanding 5000 psi. The mix designed to
meet this requirement is shown in Table 1.

APPARATUS

1. Mixing tub
2. Five wax cylinders
3. Tinius Olsen testing machine (serial number 89377,
              capacity 400,000 lb)
4. Curing room

-1-

TABLE 1: DATA AND CALCULATIONS FOR MIX

Material	Weight (lb) Per Cubic Yard	Weight (lb) Per 28.7 Pounds
Cement	775	28.7
Water	340	12.6
Fine aggregate	1,140	42.2
Coarse aggregate	1,680	62.2
Air	--	--
Total	3,995	145.7

PROCEDURE

(1) One cubic foot of concrete was mixed according to the preceding specifications. (2) The concrete was poured into five wax cylinders and tamped three times during the pouring. (3) When the concrete had hardened, the forms were removed and the concrete cylinders were placed in the curing room to be moist-cured at 70 degrees Fahrenheit. (4) In the final stage of the procedure, a Tinius Olsen testing machine was used to test the cylinders under a compressive load. One cylinder was tested at 7 days, one at 14 days, one at 21 days, and the remaining two at 28 days.

COMPRESSIVE STRENGTH

As specified, the concrete used in the Canyon Project will be subjected to a compressive stress of 5000 psi. Dur-

-2-

ing the test, however, the concrete ruptured at 3650 psi.
The designed and achieved stresses are shown in Table 2.

If the concrete was mixed and cured properly, it should
have increased in strength throughout the 28-day period. The
7-day cylinder should have been near 70% of maximum strength,
the 14- and 21-day cylinders should have shown gains, and the
final two cylinders should have reached or surpassed 5000
psi.

TABLE 2:   RESULTS

DATE TESTED	DAYS CURED	MAX. LOAD (lb)	STRESS (psi)	DESIGNED STRESS
7/6/78	7	81,000	2,840	3,400+
7/13/78	14	89,000	3,130	3,750+
7/20/78	21	104,000	3,650	3,900+
7/27/78	28	97,000	3,400	5,000+
7/27/78	28	45,000	1,580	5,000+

Calculations for amounts of water, cement, and fine and
coarse aggregates have been double-checked; no error was de-
tected in the proposed design.

CONCLUSIONS

The most likely cause for the concrete's rupture at 3650
psi is that during initial curing, prior to removal to
the curing room, the cylinders were accidentally jarred, dis-

rupting settlement of the concrete. Another possibility is that the cement was insufficiently tamped during the pouring process.

RECOMMENDATIONS

Because no problem can be found in the design, the tests should be repeated. The curing process should be closely watched, and great care should be taken during the tamping stages.

Clinton Hare

Clinton Hare

—4—

Like the informal recommendation report earlier in the chapter, this laboratory report has an inductive structure. It leads readers step by step through the investigation, the reasoning, and, finally, the conclusions and recommendations.

In some companies, writers place the conclusions and recommendations at the beginning of the report immediately following a statement of the report's purpose. This deductive format, in which the supporting evidence follows the conclusions and recommendations, allows busy readers to see the results immediately and decide for themselves whether they want to examine the data. A deductive format can be used for informal recommendation reports as well as for laboratory reports.

An informal report can also begin with an abstract, which is a concise summary of its sections. For example, the lab report could have been preceded by the following abstract:

Tests were conducted to determine whether the concrete mix proposed for the Canyon Project meets the specified compressive strength of 5000 psi.

The cement ruptured at 3650 psi on the Tinius Olsen testing machine, but double-checking indicates that calculations for the design are accurate. Human error, either in insufficiently tamping the mix, or accidentally jarring the cylinders during initial curing, is probably responsible for the failure.

The mix should be retested, with particular care being taken during tamping and curing.

In about 85 words, the readers receive the gist of the entire report, including its conclusions and recommendations. They can then read the full report if they need the details. Abstracts will be emphasized in Chapter 9 because they are commonly used in formal reports.

As this chapter has shown, memorandum forms are by far the most versatile vehicles for communicating information within a company. They are used for reports of the results of investigations as well as for handwritten messages, and Chapter 9 will show how they also function as transmittal correspondence.

9 Formal Reports

AS Chapter 8 suggested, formal reports have an important function in industry today. They provide the information that management needs in order to make decisions affecting the future of departments or entire firms. To successfully perform its function in a company's decision-making process, a formal report must communicate with many people: executive and management personnel, senior engineers, perhaps legal and financial officers, and others whose areas will be affected by the decision. The technical knowledge of these people obviously varies tremendously, but the report must serve as the main source of information for all of them. The increasing need for quick and effective communication with a wide range of readers has caused changes in the format of formal reports:

Transmittal Correspondence
Title Page
Table of Contents

CONVENTIONAL	MODERN
Introduction	Abstract
Body	Introduction
Conclusions	Body
Recommendations	Conclusions
	Recommendations
	Appendix

The modern format includes an abstract and an appendix, two elements that greatly increase a report's flexibility. An abstract is the entire report in capsule form; it serves as a miniature report within the report. The appendix, which supplements information in the body of the report, contains highly technical information. With the addition of these two elements, a report adapts itself to all readers. By examining the abstract, busy readers quickly acquire the report's essential information; if they want more detailed information on a particular aspect, they inspect the appropriate section in the body of the report. Technically informed readers study the complex information in the appendix. The appendix allows the report writer to aim the body of the report at less-informed readers.

Many industrial firms have adopted the modern report structure, also called the administrative or double-report format. Figure 9.1 illustrates its effective-

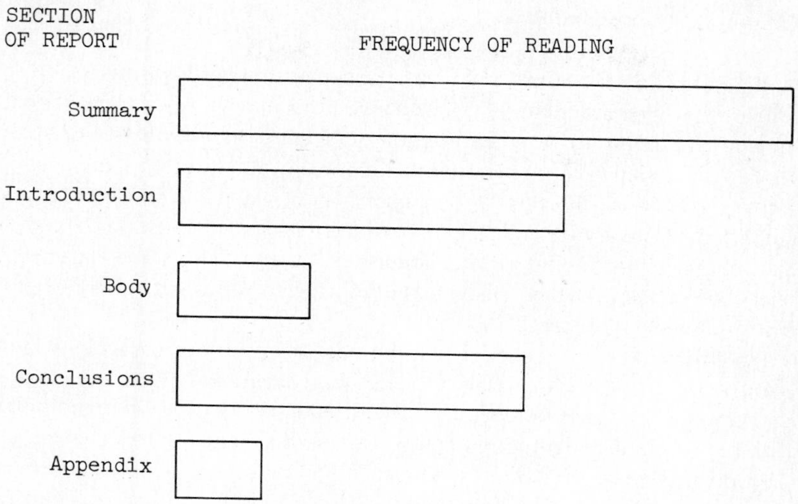

```
SECTION
OF REPORT                    FREQUENCY OF READING
```

Fig. 9.1   How managers read reports
Richard W. Dodge, "What to Report," *Westinghouse Engineer*, 22 (July–September 1962), p. 108.

ness by indicating how frequently each part of a report is read. The summary, or abstract, receives the most attention because it summarizes the entire report. Its importance seems out of proportion with the time required to write it, but this is not really the case. In effect, you work on the abstract while you prepare the body of the report; an abstract can hardly be better than the information it summarizes.

A complete formal report contains the following elements:

Transmittal correspondence
Title page
Table of contents
List of illustrations
Lists of definitions and symbols (if unavoidable)
Abstract
Introduction
Body
Conclusions
Recommendations
Bibliography (if applicable)
Appendix

Transmittal Correspondence

As the name implies, transmittal correspondence simply directs the report to someone. A memorandum (Fig. 9.2) conveys an internal, or in-firm, report. An external, or firm-to-firm, report requires a letter of transmittal (see Chapter 15, p. 252). In either form, the information remains the same. The correspondence contains (1) the title of the report, (2) a statement of when it was requested, (3) a very general statement of the report's purpose and scope, (4) an explanation of problems encountered if, for example, some data were unavailable, and (5) acknowledgment of those who were particularly helpful in assembling the report.

Transmittal correspondence should be only three or four paragraphs long because formal reports include an abstract, and saying too much in the letter or memorandum leads to repetition. The sample transmittal memorandum in Fig. 9.2 is from a student report that will be used throughout this chapter.

Traditionally, transmittal correspondence has included a final paragraph in which the writer expresses hope that the report will be satisfactory. There is nothing wrong with this, but nothing particularly good about it either; it wastes words because it tells the readers something they already know, and it simultaneously suggests the possibility that the report will not be satisfactory.

Title Page

Because the title page is among the reader's first impressions of the report, it should be well balanced and attractive. Some firms have standard title pages just as they have letterhead stationery for business letters. As Fig. 9.3 shows, title pages contain (1) the report's title, (2) the name and title of the person to whom it is addressed and the name of his or her firm, (3) the name and title of the writer, and his or her firm's name if the report is external, and (4) the date.

Table of Contents

A table of contents (Fig. 9.4) tells what the report contains and where to find it, and indicates also the organization, depth, and emphasis of the report. Special-interest readers often glance at the table of contents, examine the abstract or summary and turn to a particular section of the report. Others, after examining the conclusions and recommendations, locate certain data or calculations in the appendix.

MEMORANDUM

N. P. SCOTT MANUFACTURING COMPANY

TO: Mr. Charles Tinkham   DATE: December 5, 1978
         Manager

FROM: Thomas E. McKain
         Engineering Department

SUBJECT: FEASIBILITY REPORT ON ADJUSTABLE—SPEED DRIVES

The attached report, entitled "The Feasibility of Adjustable—Speed Drives," is submitted in accordance with your request of October 15, 1978.

The report examines possible adjustable—speed drives to incorporate on the 3—1505 lathe. The cost and capability of a direct—current drive are compared with those for an alternating—current drive.

The more feasible type of drive is recommended. Additional recommendations are made for purchasing the drive and establishing a maintenance program after its installation.

Thomas E. McKain

Thomas E. McKain

Fig. 9.2 Transmittal memorandum

FEASIBILITY REPORT ON ADJUSTABLE–SPEED DRIVE

Prepared for
Mr. Charles Tinkham
Manager, N. P. Scott Company

By
Thomas E. McKain
Engineering Department

December 5, 1978

Fig. 9.3 Title page

TABLE OF CONTENTS

Fig. 9.4   Table of contents

To assist your readers, you must maintain uniformity between items in the table of contents and headings in the report. If, in the table of contents, readers see first- and second-level headings in capital letters and third-level headings underlined, they should find precisely the same words and typographical devices in the headings of the report. Also, all items in the table of contents should be headings within the report, but not all headings in the report need to be in the table of contents. Two or three levels in the table of contents are sufficient to direct the readers, but lower-level headings may be needed within the report to achieve clarity.

Firms have various methods for numbering the pages of reports and placing items in the table of contents, but the following seems the most logical: Except for the title page, center lower-case Roman numerals (ii, iii) at the bottom of each page preceding the introduction. Starting with the first page of the introduction, center Arabic numerals at the bottom of each page. Include in the table of contents only those items following the table of contents. As Fig. 9.4 shows, indentation should be used to help distinguish various levels of headings in the table. Roman numerals often precede first-level headings, but lower-level headings are not numbered.

List of Illustrations

Illustrations are both tables and figures. Any illustrative material that is not a table is a figure, regardless of the way it is presented (drawing, graph, schematic, and so on). If illustrations appear only in the appendix, they can just be listed in that section of the table of contents. Illustrations usually appear throughout the body of the report as well as in the appendix, however, so they require a separate list of illustrations. The two kinds of illustrations should be separately numbered and listed, as shown in Fig. 9.5. Provide the number, title, and page of each illustration. The list of illustrations may be placed on the same page as the table of contents unless doing so crowds the page.

Lists of Definitions and Symbols

Traditionally, reports have included lists of definitions and symbols; however, such lists are both unrealistic and unnecessary. Highly technical terminology and symbols should not appear in the body of a report. Rather, they should be reserved for the appendix, whose readers do not need the definitions. When you must use technical terms in the body of the report, why not define

LIST OF ILLUSTRATIONS

Figures

Tables

Fig. 9.5   List of illustrations

them immediately in parentheses? Informed readers can simply skip over the definitions.

If for some reason you cannot avoid separate lists, they should be detachable so readers can place them beside the text; few readers are able to memorize quickly and retain definitions of 8 or 10 unfamiliar terms. Place symbols and definitions under separate headings, and write definitions that are parallel in structure although not necessarily in complete-sentence form.

Abstract

The abstract described here is an *informative* abstract that summarizes the entire report, serving as a mini-report. It is sometimes called an introductory summary. An informative abstract represents the entire report, allowing busy readers to grasp the report's significant information without going any farther. Another kind of abstract, the *descriptive* abstract, simply identifies areas covered in the report; it has limited value because it does not summarize the contents.

Headings generally do not appear within an abstract, but the abstract summarizes (1) the report's purpose and the problem, (2) the major facts on which the conclusions are based, (3) the conclusions, and (4) the recommendations. In Fig. 9.6, the first two paragraphs summarize the introduction of

Summary
~~ABSTRACT~~

Summary of Introduction

The purpose of this report is to determine which adjustable-speed drive, ac (alternating-current) or dc (direct-current), should be purchased for production of the 19,500 steel crankshafts ordered by General Motors.

The number 3-1505 lathe must be converted to an adjustable-speed drive because its constant speed of 1620 rpm (revolutions per minute) is too fast for making steel crankshafts. The ac and dc drives are compared according to cost and capability.

Summary of Body

Initial cost and installation of the dc drive is $122,000, versus $102,500 for the ac. The operation and maintenance of the dc is approximately $14,000 less per year, however. With all costs subtracted from the production of the 19,000 units, the dc drive would show a profit of $203,000 (12.8% increase) versus $197,000 (9.8% increase) for the ac.

The only difference in the capability of the drives is in their braking systems. The dc drive's system reduces the input power required by the lathe, but the ac's is slightly less efficient and costs about $3000 per year more to operate.

Summary of Conclusions

The ac and dc drives are both capable of doing the job, and either would increase profits. The dc, however, has a 3% greater profit potential than the ac.

Summary of Recommendations

The dc adjustable drive should be installed on the number 3-1505 lathe. It should be purchased from the General Electric Company, which has given assurance that the drive will be installed within 2 weeks after purchase. With its implementation, a preventive maintenance program should be set up to check the drive's components continually.

Fig. 9.6   Abstract

---

the report, paragraphs three and four summarize the body, and the last two paragraphs summarize the conclusions and recommendations. Try to concentrate this information into approximately 10% of the original report's length. Saying too much defeats the summary's purpose.

You should write an abstract after you have written the rest of the report. This seems repetitious, but the abstract is not repetitious for readers who have no intention of looking at the rest of the report. You should avoid technical terminology because most readers who really depend on an abstract are technically uninformed. Also, references to information in the body of the report are unnecessary; the readers have access to this information through the table of contents.

Introduction

The cliché, "What you see is what you get" applies well to the introductions of reports. An introduction prepares the readers to understand the body of the report and simultaneously keeps them from expecting something the report does not contain. Elements of an introduction are discussed in the following section and exemplified in Fig. 9.7.

INTRODUCTION

PURPOSE

The purpose of this report is to recommend that either an ac (alternating-current) or dc (direct-current) adjustable-speed drive be purchased and installed on the number 3-1505 lathe for producing steel crankshafts.

PROBLEM

The General Motors contract calls for 19,500 steel crankshafts. Number 3-1505 is the only lathe large enough to produce crankshafts, but its constant speed of 1620 revolutions per minute is good only for cutting softer nodular-iron crankshafts.

SCOPE

The ac and dc drives are compared according to their cost (break-even point, first cost, rate of capital recovery), and capability (efficiency, overload capacity, braking, availability of parts). The sections on cost and capability are preceded by background information on each type of drive.

Fig. 9.7   Introduction

STATEMENT OF PURPOSE

You can usually state the first portion of an introduction, the statement of purpose, in one or two sentences. The best way to state the purpose is directly, by saying, "The purpose of this report is . . . ." (solve whatever problem necessitated the report).

THE PROBLEM

The purpose statement leads you into the introduction's second part, a definition of the problem. Many people wonder why a definition is necessary, but it is important that the readers be completely informed about a problem before being told of its solution.

During your investigation, you may find that the problem is not what it first seemed to be; sometimes it looks complex and turns out to be fairly simple, and often the reverse is true. Because you are the investigator, you become the expert. Your readers are comparatively uninformed, and they will not understand your solution to the problem unless they view the problem as you do. To supplement the definition of an extremely complex problem, you can devote the first section in the body of the report to background information.

SCOPE

A scope statement reveals the emphasis, boundaries, and organization of a report. In feasibility reports (see Chapter 11) the scope section contains statements of the alternatives you are considering and the criteria you are using to judge the alternatives. In other kinds of reports, you identify the main sections, or topics, of the report. By listing the sections in the order they appear in the body of the report, you also indicate the report's organization. Given the type of introduction shown in Fig. 9.7, the readers reach the body of the report prepared to understand its information.

Body of the Report

The body of each kind of report contains unique elements, each of which is discussed in subsequent chapters. All formal reports, however, contain headings to clarify the parallelism or subordination of various elements and to serve as transitional devices. Headings also make reports more visually attractive and readable. Like other visual aids, they cannot be used for their own sake but should be included wherever logic permits.

First-level headings (INTRODUCTION, COST SECTION, RECOMMENDATIONS) are generally entirely capitalized and placed at the top of a new page (because of space limitations, these headings do not start a new page in the models in this book). Second-level headings are often entirely capitalized also, but as the following examples indicate, their placement varies. Third-level headings, with only the first letters of major words capitalized, are generally underlined and placed at the left margin to indicate their subordination. Fourth-level headings are capitalized and underlined like the third-level headings, but they are indented five spaces on the same line as the first sentence of the new paragraph, and they are followed by a period.

Two common methods for lettering and placing headings are shown. Any logical system may be used, however, provided the various headings appear uniformly throughout the body of the report and match their counterparts in the table of contents.

MAJOR HEADINGS	MAJOR HEADINGS
MAIN HEADINGS	MAIN HEADINGS
<u>Subheadings</u>	<u>Subheadings</u>
<u>Paragraph Headings.</u>	<u>Paragraph Headings.</u>

Conclusions

The conclusions section emphasizes the report's most significant data and ideas. <u>The readers must never be surprised by the conclusions,</u> and they will not be if you have written the report logically and clearly. Generally, you can write conclusions by summarizing the most important information in each of your report's major sections. Concise, numbered conclusions, limited to information having the greatest impact on the recommendations, are the most effective (Fig. 9.8).

    1.   The initial and installation costs of the dc drive are greater than the ac's. However, the fixed and variable costs associated with the ac drive reduce the ac's total profit—producing capability. The dc drive would produce a profit of $203,000 for the 19,000 crankshafts, which is an increase of 12.8% over the present constant—speed drive. The ac drive would increase profits only 9.8%.

    2.   The capability of the drives is very similar. The

main difference between the two is their braking systems.
The dc drive has a regenerative brake that reduces the input
power required by the lathe. The braking system of the ac
drive is just as quick as the dc but not as efficient, and it
would cost as much as $3000 per year more for input power.

Fig. 9.8 Conclusions

Recommendations

After viewing the conclusions, readers often know what the main recommendation will be. Recommendations put you on record, however, and should be as firm, clear, and concise as possible. The main recommendation usually fulfills the purpose of the report, but you should not hesitate to make further recommendations. In Fig. 9.9, the writer suggests the initiation of a maintenance program to insure that the recommended equipment will perform well. If carried too far, this becomes "hedging," but done properly it serves both you and your firm.

1. Because of its profit—making potential, capability,
and efficiency, the dc adjustable—speed drive is recommended
for installation on the number 3—1505 lathe.

2. The drive should be purchased from the General Electric Company. They have given assurance that the drive can
be installed in a maximum of 2 weeks.

3. A preventive maintenance program should be established by the machine shop. This maintenance program will
force continual checking of the drive to see that its components are performing properly.

Fig. 9.9 Recommendations

Bibliography

The bibliography, which is included when the report contains information from other sources, was discussed along with footnotes in Chapter 6. Many firms today have their own libraries, which furnish useful information for formal reports.

Appendix

Writers frequently center the word APPENDIX on a fresh page to signal its start. The appendix contains information of a subordinate, supplementary, or highly technical nature that you do not want to place in the body of the report.

Today, the trend is toward greater use of appendixes to shorten the main report. However, you must not place so much in the appendix that you fail to present significant data in the main report; to avoid this, you can often prepare simplified versions of appendix items for the main report. In any case, refer to each appendix item at the appropriate place in the body of the report.

If the appendix has major sections, the headings should be placed in the table of contents (Fig. 9.4). Illustrations in the appendix must be given titles and numbered using the numbering sequence begun in the body of the report.

Section Four

Types of Formal Reports

Section Four is organized according to a common sequence of industrial report writing. To oversimplify: An idea for a change (new system, product, or service) is presented in a *proposal*. When several proposals are submitted, a *feasibility report* may be written to determine which proposal should be implemented. As the change is being made, *progress reports* are written; and after the change is completed, new *manuals* are often needed. *Oral reporting* may reinforce or substitute for the proposal, feasibility, or progress reports in the sequence.

This overview serves as a point of departure for Section Four, but it is not meant to suggest that each formal report you write in industry will fit neatly into a sequence. The individual chapters will explain the various situations that demand each type of report.

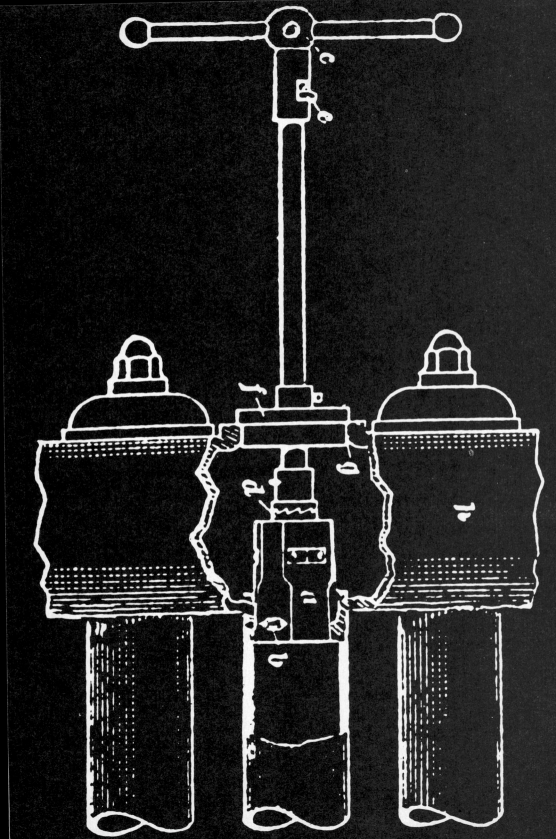

10   Proposals

A PROPOSAL presents a solution to a technical problem. There are two kinds of proposals: *external* and *internal*. The external proposal responds to another firm's request for a solution to a problem. In an internal proposal, an employee or department conveys an idea for an improvement; often unsolicited, the report goes to someone in a higher position in the firm. These two kinds of proposals have similar content, but their functions are quite different, so they will be discussed separately.

External Proposals

The survival of a firm depends on its ability to write convincing external proposals. These reports are the type that a firm writes in order to win contracts for work. An abundant source for such contracts is the government, which requests proposals for innumerable items, the most familiar being national defense weapons.

For example, the Pentagon recently asked several large aircraft companies to submit proposals for the F-16 fighter jet. Along with each invitation, the Pentagon sent specifications stating requirements for the jet. Specifications are extremely detailed and comprehensive, stating standards for minute technical items, and specifying the content, format, and deadline for the proposals.

When the Pentagon or any other large organization solicits proposals, it starts a chain reaction of proposal writing. The potential prime contractors for a new aircraft (companies such as Boeing, General Dynamics, and North American Rockwell) have the ability to design and manufacture the airframe, but that is about all. Working from the original set of specifications, each of them writes its own specifications and solicits proposals from material suppliers, equipment manufacturers, and other subcontractors who can provide such items as radar equipment and hydraulic systems. Thus, firms that specialize in particular types of equipment submit proposals and compete with each other for subcontracts.

In an intense process that often takes several months, all the firms involved attempt to write convincing proposals. As will be seen later, cost is only one of the criteria that determine who wins the contracts. The solicitor considers the quality of the product, financial status of the firm, qualifications of its personnel, and its track record for meeting deadlines. When the prime contractor finally selects its subcontractors, it compiles all the data, submits a proposal to the Pentagon or whoever initiated the whole project, and hopes for the best.

General Dynamics of Fort Worth, Texas, won the F-16 contract, and Westinghouse Corporation won contracts for the jet's radar system and electrical generating system. These contracts are worth billions of dollars; thousands of workers, from laborers to engineers and management personnel at General Dynamics, Westinghouse, and their subcontractors, are assured of jobs because of their firms' successful proposals. So many smaller firms depend on General Dynamics and other giants like it that proposal writing eventually affects a huge number of jobs. Sooner or later, most engineers become involved in the technical end of a proposal, and often the technical-writing end.

Not all proposals are written for national defense items. Proposals are also common in state and local governments, public agencies, education, and industry. Competitive bidding through proposal writing is a basic part of the free-enterprise system.

ELEMENTS OF AN EXTERNAL PROPOSAL

The three main parts of a proposal are its technical, managerial, and financial sections, each of which amounts to a proposal in itself. None of these parts can be considered more important than the others. Cost has great importance, but the fact that the price is right becomes meaningless if the product is not, and both the price and the product become insignificant if a firm misses a deadline because of faulty management.

The difference between a winning and losing proposal is often an intangible: confidence. The tone of a proposal must be positive to convince readers that the company can be trusted to produce for them. Injecting confidence into a proposal presents more problems than might be readily apparent. The firm writing the proposal, like the firm requesting it, confronts the unknown. When its specifications were sent to bidders, the F-16 was nothing but an idea, some words and drawings on pieces of paper. So were its hydraulic systems and other components. To project into the future and convey a positive attitude, you must have confidence in your own company and your ability as a writer.

Technical Section   A proposal's technical section begins by stating the problem to be solved. This seems unnecessary, but the firm must clearly demonstrate that it understands what the solicitor expects. Then, the firm describes its approach to the problem and presents a design for the product if one is needed. Sometimes the firm offers alternative methods for solving the problem and invites the solicitor to select one.

Management Proposal The management proposal outlines the chain of command that will be followed during work on the project. The solicitor wants to know that the project will be given top priority, and the proposal writer must respond by explaining what positions and levels of management will be responsible for the success of the project. For added assurance, this portion of the proposal thoroughly describes how quality, costs, and schedules will be controlled. Finally, the firm includes a statement concerning its facilities, finances, and previous contracts.

Cost Section The cost section provides a breakdown of the costs for every item in the proposal. Cost also involves the unknown because the firm must look into the future and estimate changes that may occur in the cost of labor, parts, and material during its work on the project.

   After all the proposals are sent to the firm that requested them, they are studied by a team of evaluators, some of whom helped write the original specifications. First, the evaluators rate the proposals' technical and management sections without consideration of the cost data. A firm that submits an inadequate design or fails to communicate its design effectively loses the contract no matter how low its price. Then, evaluators examine the cost figures and select the best overall proposal.

Internal Proposals

All projects or changes begin somewhere, and that somewhere is usually on paper. To receive attention, an employee's idea generally has to be put in writing. Once on paper, the idea becomes an internal proposal that will be evaluated by those in supervisory and management positions. Needless to say, departments have grown and careers have been launched by proposals. The person who comes up with an idea for a project, and effectively communicates that idea, gains recognition and becomes a logical choice for a role in the implementation of the project.

   An outline for an internal proposal is shown, with the sections in the body of the report linked to their counterparts in the external proposal:

Title Page
Table of Contents
List of Illustrations
Abstract

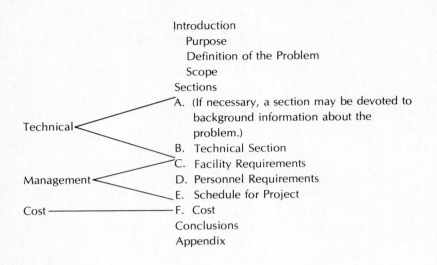

Introduction
   Purpose
   Definition of the Problem
   Scope
Sections
A. (If necessary, a section may be devoted to
   background information about the
   problem.)
B. Technical Section
C. Facility Requirements
D. Personnel Requirements
E. Schedule for Project
F. Cost
Conclusions
Appendix

Technical
Management
Cost

   Chapter 9 described a formal report's title page, table of contents, list of
illustrations, abstract, and appendix. The following discussion concentrates
on the introduction and body of an internal proposal.

INTRODUCTION

The introduction to an internal proposal contains a statement of the report's
purpose, a detailed explanation of the problem, and a concise statement of
scope.

Purpose   A proposal's purpose is to present a solution to a particular problem.
In stating the purpose, you should identify the problem but reserve amplification of it for the next section of the introduction.

Problem   The definition of the problem is the most important portion of a
proposal's introduction. As the writer of an unsolicited proposal, you must
demonstrate that a serious problem exists in order to justify the existence of
your proposal. Even if the proposal has been solicited, your investigation
should have given you greater insight into the problem than your readers
possess. In either case, your definition must be clearly understood because
the body of the report will attack the problem as you view it. As the preceding
outline indicates, you may place a section of background information in the
body of the report. The section may include a detailed description of the

existing situation (the problem) or a discussion of previous approaches to the problem.

Scope In the statement of scope, name the report's major sections in the order they will be presented. You thereby identify the boundary of the report and clarify its organization.

SECTIONS

The sections in the body of a proposal vary according to the type of problem being attacked; you must determine the content and structure that will best serve your proposal. However, the five sections explained on the following pages should at least be considered for inclusion in every internal proposal.

Technical Section The technical section is generally the longest section of an internal proposal. In it, you provide a detailed explanation of your technical solution to the problem. For example, suppose that your company is unable to keep up with its orders because its assembly line is too slow. To solve the problem, you propose that a new and faster mechanism be purchased for use on the assembly line. In the technical section of such a proposal, you would describe the mechanism you want purchased, explain that it would be compatible with other equipment on the assembly line, and emphasize its ability to solve the company's problem by speeding up the assembly line.

Throughout a proposal, and particularly throughout its technical section, remember that your readers have not asked for the report and are not familiar with its contents. You must present your ideas very clearly for them to be understood and accepted.

Facility-requirements Section In internal proposals, *facilities* generally refers to the physical capabilities of the firm. The availability of facilities plays an important role in proposals involving such things as the creation of a new product, enlargement of a department, or installation of a larger, more modern system. If the suggested change can be implemented without additional space, that is an advantage for the proposal. If, however, the proposal requires an expansion of present facilities, you must describe the physical alterations. Rather than state their cost in this section, wait until the cost section, where the readers can see expenditures in the proper perspective. It is to be hoped that the detailed information that precedes the cost section will persuade the readers to accept the cost figures.

Personnel-requirements Section   Many proposals advocate the installation of new equipment. Such a report's personnel-requirements section can be divided into two sections, one concerning personnel necessary to complete the project, and the other involving personnel needed to operate the equipment after installation. In the first portion, keep in mind that many firms today employ people with wide experience and diverse specialties; investigate your firm's potential and, if possible, suggest that the project be accomplished without costly outside help. The second portion deals with personnel required after completion of the project. If additional people must be hired to operate the equipment, you should state so explicitly. If retraining the personnel is necessary, you should explain the length and type of training. Again, you should withhold the costs for personnel until the cost section.

Schedule Section   Always assume that your proposal will be accepted, anticipate difficulties that will arise during its implementation, and attack them. A work schedule is one of those difficulties. Eventually, someone must prepare a schedule for the project, and it might as well be you, particularly if you want to show how thoroughly you have planned the project. In the preceding example concerning a conveyor proposal, the writer should be capable of suggesting a reasonable date for the project to start, one that will allow its completion with minimum disruption of the firm's production schedule. The writer should also estimate the time required for each phase of the project. If the proposal includes a detailed schedule, several potential management headaches are avoided, and the proposal has a better chance for acceptance.

Cost Section   The cost section is the most important part of an internal proposal. It contains an itemized list of prices for everything involved in the project, but even more important, it explains how the money will be regained. In industry, an idea is as good as it is profitable; firms can usually find the money if they are assured that they can regain it and show a profit. Therefore, present clear and convincing evidence that your proposal has profit potential and should be acted on. The time-to-recover method discussed in Chapter 11 is one effective way to indicate eventual profit, but any logical presentation will be effective.

CONCLUSIONS

The conclusion of a proposal is essentially an inducement to action. There are three possible reasons that a problem has not been acted on: people with the power to correct the problem (1) are not aware of the problem, (2) *are* aware of the problem but do not care, (3) are aware of the problem and *do* care but either do not know how to correct it or think that correcting it will cost too much. In the body of the report, you have anticipated objections to your proposal and attacked them. In effect, you have attempted to convince the readers that solving the problem presents no problems. In the conclusion, re-emphasize the advantages and strong points of your proposal.

The two student proposals that follow should give you some ideas for topics and for organizing your report. To save space in the bodies of the model reports, each main heading does not start a new page as it normally would.

Models

PROPOSAL FOR
PROTECTION AGAINST SPREAD OF ALPHA RADIOACTIVITY

Prepared for
Mr. Donald Smith
Industrial Hygiene Department
Gawthrop Laboratories

By
Paul Markovich
Engineering Assistant, Plutonium Laboratory

October 29, 1978

TABLE OF CONTENTS

LIST OF ILLUSTRATIONS

ABSTRACT

Alpha radioactivity has often been spread over laboratory areas through the carelessness of people working in the area. Although detectors are available, they are not always used, causing the radioactivity on people to be inadvertently spread. By means of an electric-eye alarm scanning the laboratory's entrance, those who fail to use the radioactivity monitor will set off an alarm that can be stopped only when they use the monitor to check themselves. The alarm will warn personnel when they are entering as well as leaving the laboratory.

Our Electronics Division can adapt the present system to include an electric eye. Costs for this virtually foolproof system will total approximately $870, including a new monitoring unit and 1 year's maintenance on the entire system.

INTRODUCTION

PURPOSE

This report proposes a system for stopping the accidental spread of alpha radioactivity by people entering and leaving the plutonium laboratory.

PROBLEM

For years, radioactivity has occasionally been spread throughout laboratory areas that contain plutonium. There are various causes for these accidents, but the major problem seems to stem from individuals failing to monitor themselves when working in these areas. Currently, personnel are expected always to remember to check themselves on the alpha-radiation monitor, which is placed by the door of the laboratory. The inefficiency of this system has been proved on numerous occasions when alpha activity has been spread.

Alpha activity, which is the major hazard when working with plutonium, can go undetected and be spread quickly unless complete self-surveillance is undertaken. The chance of danger to the carrier, the shutdown of operations until the cleanup is complete, and the labor of reclamation crews make a more formidable monitoring system mandatory.

SCOPE

In this report, a description of the proposed system for monitoring alpha radiation is presented, followed by sections concerning equipment available, personnel available, costs, and finally, conclusions.

PROPOSED SYSTEM

This report proposes the installation of a foolproof method of surveillance for all persons leaving or entering the laboratory. By means of an electric—eye alarm system scanning the entrance to the laboratory, all personnel will be reminded immediately that they are leaving or entering the area without checking themselves. If people pass into or out of the laboratory without properly checking themselves on the monitor, the electric eye will activate an alarm bell, re— minding them that they have forgotten. They will have to use the monitor to deactivate the alarm system.

With the proposed system, a monitor will be placed imme— diately outside the laboratory door. As a worker depresses the probes on the monitor, a microswitch in the probe will de—energize the electric—eye system, which will be just be—

-2-

yond the monitor. The electric eye will remain de-energized for about 10 seconds to allow the person time to pass through its beam area. Once the 10 seconds are up, the electric-eye system will become activated again.

People entering the laboratory will also be required to check themselves. This precaution is taken so that radiation from outside sources cannot be tracked into the laboratory. A monitor will be set up on the outside of the electric-eye beam, and anyone entering the laboratory without de-energizing the beam will also cause the alarm to sound. Figure 1 shows where the beam and monitors will be placed.

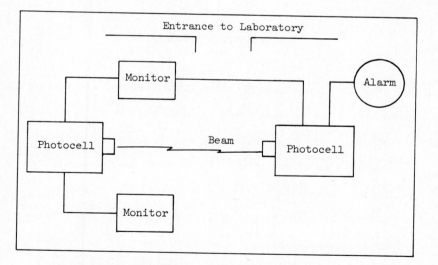

Fig. 1 DIAGRAM OF ELECTRIC-EYE SYSTEM

EQUIPMENT AVAILABLE

With the exception of one radiation monitor, which will have to be purchased, the existing equipment for detection of activity can be quickly and cheaply modified to meet the requirements of the proposed system. The major portion of the electric-eye alarm system can be readily constructed by our Electronics Division. The electronics staff can also provide parts necessary for maintaining the monitors and electric-eye equipment. The new, transistorized units will consume less space than the present system requires.

PERSONNEL AVAILABLE

Construction of the proposed monitoring and electric-eye alarm system can be undertaken by personnel in the Electronics Division, who are experienced in designing and miniaturizing this type of equipment. The mechanical and electrical equipment can also be maintained by electronics personnel. Periodic maintenance checks will be provided to insure optimum performance of the system at all times. If the system ever needs updating, the electronics staff will be able to make the necessary modifications.

-4-

COSTS

With the budgets in all laboratories somewhat limited, the costs for converting available equipment become important. However, the availability of most of the major components and the cooperation of the Electronics Division will hold the costs for this improvement to a minimum.

The major costs, at this time, appear to be those of the new radiation monitor (approximately $350) and the installation of the electric-eye equipment (approximately $240). Total expenses for the system, including maintenance costs for 1 year, are estimated at $870. Table 1 contains a complete breakdown of estimated costs.

TABLE 1: COST ESTIMATES

Item	Cost
Equipment	
2 Photocells —	$100.00
1 Alpha-radiation monitor — — — — — — — — — — — — — —	350.00
Total — — — — — — — — — — — — — — — — — — —	$450.00
Installation labor	
2 Workers at $15.00/hour for 8 hours — — — — — — — —	$240.00
Yearly maintenance	
1 Worker at $15.00/hour for 12 hours (1 hour/month)	$180.00
TOTAL ESTIMATED COSTS — — — — — — — — — — — —	$870.00

CONCLUSIONS

1. The present method of checking alpha radiation on people
entering and leaving the plutonium laboratory does not work
satisfactorily. The protection of each person has been left
up to the person, and by either forgetfulness or complacency,
monitoring has sometimes been overlooked.

2. The proposed electric—eye alarm provides an almost fool—
proof system for monitoring personnel as they enter and leave
an active area. To turn off the alarm, personnel must check
themselves.

3. The safety feature of this alarm system would be a very
important asset to the plutonium facility. The safety of
people in the laboratories, and the elimination of costly
cleanup caused by the spread of radioactivity, make this
alarm system a necessity.

PROPOSAL FOR
INSTALLATION OF A POWER GENERATOR

Prepared for
David J. Boliker
Manager of Manufacturing
Lansing Steel Company

By
Robert J. Smith
Acting Project Engineer

January 15, 1978

TABLE OF CONTENTS

ABSTRACT

The purpose of this report is to propose a back-up generating plant at the new Lansing facility.

The problems that might result from an electrical power failure at the plant involve both personnel and facilities. Employee injuries could result from moving about with no lighting in the plant, being trapped inside closed tanks, and being struck by items falling from cranes that are operated by vacuum lifts or battery magnets. The plant itself would be in danger of damage from surface-drainage water, fire, and the freezing of pipes and heaters.

A surplus generator is available from our Field Force Division and could be installed for approximately $4060, including labor and material. It can be installed in the powerhouse's air-compressor room during the month of June, and its operation will require minimum training of personnel. The generator can meet all the plant's emergency requirements and do so very economically.

INTRODUCTION

PURPOSE

This report proposes the installation of an on—site
electrical generating plant at the new manufacturing facility
in Lansing.

PROBLEM

At present, a power failure would result in the following unsafe conditions for the employees: (1) plant lighting
would be lost, and employees could not move safely within the
plant; (2) employees working inside closed tanks could not
get out until the tank could be rotated by an electric motor;
(3) employees could be injured by items falling from the
three overhead cranes that hold their loads by vacuum lifts
or battery magnets and must set them down within 10 minutes
after power failure.

A power failure would also cause damage to the plant:
(1) the sump pump for surface—drainage water would not operate in the main office building and the basement of the grit—
blast area, causing extensive damage to equipment; (2) the
fire alarm systems and the main fire pump would not function,
and a fire could not be contained; (3) during winter, loss of
the main heating boilers would cause water lines and fire—

loop feedlines to freeze, break, flood areas of the plant, and ruin equipment.

SCOPE

In Section I, this report describes the installation of the generator. This is followed by sections on lighting and equipment changes, the schedule for the project, personnel training, and cost. At the end of the report, conclusions are presented.

SECTION I: INSTALLATION OF THE GENERATOR

A back-up generator is readily available for this project. One was purchased by the Field Force Division in 1970 for use as a standby unit at building sites. Because it was used only while other units were being repaired, the generator has operated for just 565 hours. Similar units have operated in excess of 15,000 hours, so the generator is almost new. It was built by Westinghouse and has a General Motors diesel engine.

The generator can be located in the air-compressor room of the powerhouse. This location will allow easy hookup to the main utility substation. The floor of the powerhouse is sufficient to hold the weight of the generator and to retain

bolts needed to anchor the unit. The start—up fuel tank, lo—
cated near the generator, will be used to operate the diesel
engine.

Siding along the powerhouse's east wall must be removed
to allow entrance of the generator; a louvered panel will re—
place the siding and allow cooling air to circulate around
the generator. Installation of the generator will require
drilling holes in the floor to mount the anchor bolts. A
rack must be built to hold the batteries for starting the
diesel engine. An electric load—transfer panel will be as—
sembled and wired into the existing substation, and a fuel
line will be connected to the 2000—gallon fuel tank. Hook—
ing the generator to the substation requires installation of
three 460—volt cables between the powerhouse and the trans—
former. This can be accomplished by maintenance personnel.

SECTION II: LIGHTING AND EQUIPMENT CHANGES

In order to stay within the capacity of the generator
and still provide the power needed to maintain necessary
lighting and motor power, some electrical wiring changes must
be made in the plant. Lighting circuits now operated by me—
chanical switches will have to be changed to electricity.
This change means that if the electric power fails, the

-3-

lighting will not turn on after auxiliary power is established, thus removing electrical loads during power failure.
However, the night lights, which will turn on automatically
with the start of auxiliary power, will allow employees to
leave the plant safely.

The rotoblast power supply will have to be changed to a
manual-start mode. This conversion requires the addition of
one relay to the panel board and allows personnel operating
the cranes to lower suspended loads safely to the floor.

Sump pumps used for the removal of drainage water from
the office building and basement of the gritblast will automatically return to operation when auxiliary power is supplied. No changes will be required in the existing circuits
to these pumps.

The fire alarm will function as soon as power is trans-
ferred to the generator. If a fire should start, the alarm
would be sounded immediately. At that time, all electrical
power would be turned off, including the night-light circuitry. This is necessary because the fire pump's motor
taxes the generator so much that very little electricity is
available.

SECTION III: SCHEDULE FOR THE PROJECT

PHASE ONE

Phase one will be completed during the normal work week, using plant personnel. This phase will include wiring the rotoblast, changing the lighting, installing anchor bolts, installing the fuel line, building and installing battery racks, and wiring the generator to the substation.

PHASE TWO

Phase two will involve Saturday work because two shifts of four workers each will be required, and these workers cannot be spared from duty during weekdays. Work will consist of removing the siding on the powerhouse wall, moving the generator into the powerhouse, installing it on its anchor bolts, and replacing the siding with a louvered panel.

PHASE THREE

Phase three will be accomplished on a Sunday when power can be turned off in the plant. Work will consist of hooking cables into the outside substation, hooking the fuel line to the engine, and connecting the power and battery cables to the generator. The power will be off for only half an hour, and no special precautions should be necessary because no operating personnel will be in the plant.

This project can begin 4 weeks before the start of plant shutdown, which is scheduled to begin July 7. Phase one can start June 12 and be completed by June 21. Phase two can be done Saturday, June 22, and phase three can be completed Sunday, June 30.

SECTION IV: PERSONNEL TRAINING

No personnel training will be needed for starting the diesel generator. The control panel installed in the power-house will automatically start the engine in the event of a power failure and transfer the plant's reduced electrical load to the generator.

Operation of the fire pump from auxiliary power requires the manual operation of a start switch. All members of the plant fire team are familiar with this operation.

SUPERVISION

All shop supervisors will be instructed about the limitations of the auxiliary power. These supervisors will tell their personnel which equipment should be operated to assure a safe shutdown. The supervisors have enough experience in their areas to know what needs to be done. All instructions to supervisors will be covered by an interoffice memorandum when the auxiliary generator is operational.

MAINTENANCE

Routine maintenance on the engine and generator, using plant maintenance personnel who are familiar with this type of equipment, will be scheduled. The generator will be tested once per month to insure its availability for an emergency.

SECTION V: COST

The generator, which cost $14,657 in 1963, is now worth approximately $4330 and is available at no cost.

Following are estimated costs for installing the generator, which include labor to build and install the necessary equipment, and costs for material.

Labor rates for plant personnel are based on the latest figures for this plant. Straight-time labor is $7.00 per hour, time and a half is $9.25 per hour, and double time is $11.50 per hour:

Material	$2,450.00
Labor	1,240.00
	$3,690.00
Contingency (10%)	370.00
TOTAL COST FOR PROJECT	$4,060.00

—7—

CONCLUSIONS

1. Our new manufacturing facility in Lansing represents
an investment of $7 million and employs 300 people. Any in-
terruption of electric power would create a total blackout
and completely halt the equipment. This would result in very
unsafe conditions for employees trying to move within the
plant. Damage to equipment and spoilage of materials would
also result if the power failure were prolonged.

2. Considering the safety of the employees, $4060 for a
back-up generator is an extremely reasonable investment. In
addition, the facilities and equipment that might be damaged
during a power failure would cost more than installation of
the generator.

Writing Assignment

Write a 1,000-word, problem-oriented formal proposal aimed at an uninformed reader. For your topic, either identify a real technical problem you are familiar with or create a hypothetical but realistic problem situation of your own. Define the problem clearly in your introduction. For the body of your report, consider each of the items in the outline early in this chapter. Base the proposal on factual data and make it convincing.

11　Feasibility Reports

IN industry, change is a sign of stability; firms must change in order to remain competitive. Fierce competition with other firms demands that companies constantly create new products, modify old ones, and develop faster, more economical methods of producing their products. Because of the emphasis on innovation, change is not only a sign of stability, it is the most stable thing in industry; if firms can be certain of nothing else, they know that tomorrow will bring change.

Ideas for change are molded into proposals and submitted for evaluation. When one proposal is presented, the company must decide whether to act on it; if several are presented, the company must select the best one. Separating the good ideas from the bad, and the practical ones from the spectacular, requires extremely thorough investigation. Few firms can afford to implement an idea that does not produce the expected results.

The job of judging proposed ideas and recommending whether they should be acted on is accomplished in feasibility reports. In industry, *feasibility* implies more than *possibility*; a feasible idea is one supported by evidence that it will succeed. A department that wants to make expensive changes must often demonstrate the feasibility of its suggestions. For larger matters—such as expansion, development of a new product, or purchase of a new system— a team investigates all aspects of the proposed change before making a recommendation.

Selection of Criteria

The selection of criteria, the most important elements in a feasibility report, occurs during the gathering of data. Criteria are the standards by which the proposals are judged; therefore, they have tremendous impact on the final recommendation. Criteria vary according to the type of feasibility problem, as the following hypothetical situations indicate:

Problem: a city needs additional airport facilities
Alternatives
1. expand the present airport
2. build a new airport at site A
3. build a new airport at site B

Criteria
1. cost
2. capability
   a. land and air space available
   b. access to highways and commercial centers
   c. ecological effects (air, water, noise pollution)

Problem: a firm needs a data-processing system
Alternatives: various-sized computer systems
Criteria
1. cost
   a. initial
   b. operating
   c. maintenance
2. capability
   a. to meet firm's present requirements
   b. to meet requirements if firm expands

Problem: the site for a branch plant must be selected
Alternatives: proposed sites in various regions of the country
Criteria
1. cost
   a. construction
   b. transportation
   c. labor
   d. taxes
2. capability of local utilities

Good feasibility reports include all pertinent criteria, but as the examples suggest, the criteria can often be grouped into the major categories of cost and capability. These criteria do not apply to all feasibility problems, but an effort should always be made to limit the number of major criteria; doing so helps organize both the investigative and writing aspects of a feasibility study.

Report Structure

Because of the many aspects of feasibility problems and the large number of potential readers, feasibility reports invariably demand a formal report structure. The following format suffices for virtually any feasibility report:

Title Page
Table of Contents
List of Illustrations
Abstract
Introduction
  Purpose
  Definition of the Problem
  Scope
    Alternatives
    Criteria
Sections
  A. (If necessary, a section may be devoted to additional background and
     introductory information)
  B. Presentation and Interpretation of Data for the First Criterion
  C. Presentation and Interpretation of Data for the Second Criterion
  D, E, etc.: Same as for other criteria
Conclusions
Recommendations
Appendix

Chapter 9 thoroughly described the elements of a formal report. Portions of a feasibility report appeared throughout that chapter to exemplify the title page, table of contents, list of illustrations, introduction, conclusions, and recommendations of a formal report. Therefore, this chapter concentrates on the presentation and interpretation of data in the main body, or sections, of a feasibility report.

Presentation and Interpretation of Data

The body of a good feasibility report must be structured according to criteria, with each criterion receiving a major heading. This format is identical to the one prescribed for recommendation reports in Chapter 5. If a report is *not* organized according to criteria, it must be organized according to alternatives, resulting in a very weak report that prevents rather than enhances comparison. Such a report tells all about the first alternative, then all about the second alternative, and so on. Related data are actually separated instead of being together as they are in a report structured according to criteria. For example, cost data for the alternatives are often several pages apart.

  Needless to say, such a structure forces the reader to do the comparing that the report writer should have been doing all along. A report structured according to criteria is not easy to write but it is much easier to read and understand.

Beneath each heading in the body of a feasibility report, you should define the criterion named in the heading. Then you present and interpret data for all the alternatives, judging them according to the particular criterion. Repeat this process for each criterion in the report.

COST SECTION

In addition to initial costs, the cost section of a feasibility report often includes estimated expenses for installation, operation, and maintenance. The report excerpted in Chapter 9 attempts to determine whether an ac (alternating-current) mechanism is more feasible than a dc (direct-current) mechanism for manufacturing crankshafts. The writer of that report presents the cost data in the following table:

TABLE 1: COST DATA

ITEM	AC DRIVE	DC DRIVE
FIRST COSTS		
SPEED DRIVE	$76,000	$110,000
AIR DAMPER	1,500	----
TACHOMETER	2,000	----
INSTALLATION	23,000	12,000
TOTAL	$102,500	$122,000
OPERATIONAL COSTS		
MATERIAL AND LABOR	$522,175	$520,620
PRODUCTION LOSSES	325	130
TOTAL	$522,500	$520,750
MAINTENANCE COSTS		
PREVENTIVE	----	$1,950
REPAIR	$20,000	5,000
TOTAL	$20,000	$6,950

Interpreting the data in this table, the writer explains that the operational costs are based on the production of 6500 crankshafts per year. The readers are then told that operational and maintenance costs, unlike first costs, are fixed; in other words, the firm would incur operational and maintenance costs yearly but would pay the first costs only once. The writer emphasizes that the ac drive's higher operational and maintenance costs would eventually exceed the dc drive's higher first costs. For this reason, the writer states that the dc unit has greater profit potential.

If your cost section or any other section in the body of your feasibility report is lengthy and complex, you can end it with a short summary. Summaries, however, can become repetitious and should be limited to sections whose complexity demands them.

CAPABILITY SECTION

Just as various cost criteria can be combined into one section of the report, the rest of the criteria often fall into the category of capability. As the earlier hypothetical situations suggested, these criteria include items that are difficult to translate into accurate cost data, such as ecological effects or the ability of a system to meet requirements in the event of expansion. Noise pollution, which could result in negative public relations for the airport, must be considered during the selection of an airport site even though no one knows how it might affect profits. Often, a system's ability to handle expansion must be examined even though no one can predict the extent of future expansion. Because such criteria lack accurate data, they must be placed in the capability section rather than the cost section of a report.

Many "capability" items do not lend themselves to the presentation of data in tables, usually because precise data are not available. In such situations, you must simply describe their potential effects as thoroughly as possible.

COST AND CAPABILITY COMBINED

An alternative method for determining feasibility is to evaluate the alternatives according to time to recover, a concept that is becoming increasingly common in industry. This method applies only when all the data, including capability data, are convertible into cost figures. It requires extremely careful projections into the future, and its result is a report with only one criterion, cost. The process for figuring time to recover is shown in the following summary of a feasibility report.

The St. Louis warehouse of a steel company does not have coil-processing equipment and can produce only flat-sheet products. Its annual sales are 2260 tons, 1.3% of the area's total market of 198,800 tons. The firm loses a large percentage of the market because approximately 75% of the sheet products sold in the area are coiled rather than flat. The area's growing market is forecast at 277,000 tons, but the warehouse will be able to compete for only a small portion of it.

Faced with this situation, the steel company decides to study the feasibility of installing coil-processing equipment. The final decision will be largely based on the time necessary for the firm to recover the money invested in the equipment, commonly called time to recover or years to recover. If the investment will pay for itself within a reasonable time, it will be considered highly feasible, and the change will be made. Figures for the cost of the new coil equipment are shown in the following table.

```
52-Inch Coil Slitting and Banding Line     $211,850
                         Installation         71,000
               1 Jib Crane (6-ton)            8,500
                                 Base       $291,350
                  Contingency (4.8%)          14,650
                                Total       $306,000


             60-Inch Coil Cut-to-Length Line  $182,850
                         Installation          59,500
               1 Jib Crane (6-ton)             8,500
                                 Base       $250,850
                  Contingency (4.8%)           13,150
                                Total       $264,000
                          Grand Total                    $570,000
```

The question now becomes "How soon can a $570,000 expenditure be recovered?" The answer necessitates thorough estimates of the potential market and projections of the firm's share of that market:

```
Added Sales--15,600 Tons at $182/Ton                     $2,839,200
                 Less Net Material    $2,189,900
                    Selling Expense      348,800
                       Property Tax        5,700
                          Sales Tax        8,000
                   Depreciation/Year      22,800
                             Total    $2,575,200
         Net Profit before Federal Tax                     $272,000
                     after Federal Tax                     $124,000
```

Recovery of $570,000 capital at 3 year's penetration--4.9 years

Although an entire time-to-recover evaluation includes additional items, the figures show that the company expects, after 3 years of selling coil-processed steel ("penetration of the market"), to sell an additional 15,600 tons per year. Subtracting the expenses for material, marketing, taxes, and depreciation, the writer shows that the new equipment will earn $124,500 during its third year. The original $570,000 expenditure will be recovered in 4.9 years.

Model

FEASIBILITY REPORT:

LEASING OR PURCHASING A CRANE

Prepared for
Mr. George Strauch, President
Southern Construction Company

By
William S. Smolen
Assistant Manager

October 9, 1978

TABLE OF CONTENTS

LIST OF ILLUSTRATIONS

ABSTRACT

The purpose of this report is to determine whether it would be more feasible for Southern Construction Company to purchase an Anderson crane or to obtain a 10-year lease on one.

Southern will begin work on the steel superstructure of a 75-story office building in February 1979. An Anderson crane is required to complete this project and fulfill contracts Southern hopes to win in the future. The feasibility of purchasing versus leasing the crane will be judged according to cost and requirements for supervision and maintenance.

The cost of leasing the crane would be $253,400. Borrowing money and purchasing the crane outright would cost $229,590, and an additional $2000 would be required to train supervisory personnel. Thus, the total cost for purchasing would be $21,810 less than leasing. The schedule of payments also favors purchasing: purchase payments during the first 5 years would total approximately $106,000, versus approximately $230,000 during the same period if the crane were leased.

Because purchase of the Anderson crane would cost less and offer a better schedule of payments, it is more feasible than leasing the crane.

INTRODUCTION

PURPOSE

The purpose of this report is to determine whether Southern Construction Company should purchase or lease an Anderson crane.

PROBLEM

In September, Southern won a contract to construct the steel superstructure of a 75-story office building. The project, scheduled to start in February 1979, is the largest the company has won. To complete this project and others Southern hopes to win contracts for, an Anderson crane must be acquired.

SCOPE

Southern has narrowed its alternatives to either purchasing the crane, which means borrowing the money from the First National Bank, or leasing the crane from Mitchell Leasing Company. The feasibility of these alternatives will be determined according to the criteria of (1) cost and (2) supervision and maintenance requirements.

-1-

COST

LEASING

Mitchell Leasing Company has proposed that Southern obtain a 10-year lease on an Anderson crane. For the first 5
years, Mitchell would charge $20 per month per $1000 of acquisition cost. The crane's acquisition cost is $400,000, so
rent would be $46,080 per year for the first 5 years. Mitch-
ell's fee is $2 per month per $1000 of acquisition for the second 5-year period, and payments would be $4608 per year.
These figures take into account the 52% tax shield covering
the crane's rent, as shown in Table 1. The total 10-year
cost for renting the crane would be $253,440.

TABLE 1: LEASING COST

Years	Rent Per Year	Less Tax Shield (52%)	Annual After-Tax Cost
Fiscal 1980-84	$96,000	$49,920	$ 46,080
Fiscal 1985-89	9,600	4,992	4,608
			$ 50,688
		x 5 =	$253,440

PURCHASING

To purchase an Anderson crane outright, Southern would
have to borrow the $400,000 from the First National Bank. The

-2-

bank has proposed a 10-year loan bearing interest at 6%, payable monthly. The loan would be secured by a mortgage on the heavy equipment Southern now owns. As shown in Table 2, the net cost of purchasing the crane would be $229,590. This cost takes into account the 52% tax shield on interest, the 52% shield resulting from the crane's depreciation, and the crane's book value after 10 years.

TABLE 2: PURCHASING COST

Price of crane		$400,000
Total interest	$132,000	
Less tax shield (52% of Interest)	68,640	
Net interest cost		63,360
Cost		$463,360
Less depreciation tax shield (52% x $346,312 10-year depreciation)	180,082	
Less book value	53,688	
Net cost		$229,590

SUPERVISION AND MAINTENANCE

LEASING

Mitchell Leasing Company's fee covers the temporary services of a field engineer. These services are mandatory

-3-

in all Mitchell's leasing agreements. The engineer supervises the crane's erection and the installation of additional
sections as construction proceeds. During the first year of
operation, the engineer inspects periodically to see that the
crane is properly lubricated and that pulleys and cables are
regularly checked for wear. He or she also supervises dismantling when the first project is completed.

PURCHASING

Although several Southern employees are qualified to operate the Anderson crane, none of them has experience in
erecting and dismantling it. However, two of Southern's senior field engineers have directed these operations on smaller
cranes for more than 10 years and could quickly become familiar with the Anderson. Their training would take approximately 1 month and cost a maximum of $2000. Lubrication and
inspection of the crane is similar to that of smaller crnaes
and could be handled adequately by Southern personnel.

CONCLUSIONS

Table 3 shows that purchasing the crane would cost
$23,810 less than leasing it. When $2000 for training supervisory personnel is subtracted from this figure, the difference

-4-

becomes $21,810. The schedule of payments also favors pur-
chasing the crane: leasing payments are $46,080 for each of
the first 5 years before dropping to $4608 during the last 5;
purchasing payments rise gradually from $9920 to $29,873 during
the first 5 years and reach a maximum of $41,152 during the
10th year. In all probability, Southern will be better able
to meet high payments during the second 5 years.

TABLE 3: SUMMARY OF COSTS

	Purchasing	Leasing
After-tax cost	$463,360	$253,400
Less depreciation tax shield	180,082	----
	$283,278	$253,400
Book value	53,688	----
Net cost	$229,590	$253,400

RECOMMENDATIONS

Because of purchasing's lower total cost and favorable
schedule of payments, it is recommended that Southern borrow
$400,000 from the First National Bank to purchase the Anderson
crane.

-5-

Writing Assignment

Write a 1000-word problem-oriented formal feasibility report aimed at an uninformed reader. Your problem situation may be either real or hypothetical, but your data should be factual. Follow the outline presented in this chapter and evaluate two possible solutions to the problem. Clearly define the problem in your introduction and carefully select your criteria for evaluating the alternatives. Structure your report according to criteria rather than according to alternatives.

12   Progress Reports

PROGRESS reports are written to inform management about the status of a project. Submitted regularly throughout the life of the project, they let the readers know whether work is progressing satisfactorily, which often means within the project's budget and time limitations. Contracts won through external proposals invariably require that winning firms provide regular reports on progress to their clients. On the internal level, departments furnish reports on a wide range of projects involving research, construction, and installation.

Progress reports are submitted at regular periods but differ from periodic reports such as weekly sales reports and monthly production reports. Unlike progress reports, periodic reports are continual, because the data they report never stops accumulating. Most firms provide standard, fill-in-the-blank forms to simplify the writing of periodic reports.

Also related to, but different from, progress reports are interim reports. The word *interim* means temporary in this context. Workers near the completion of a project are often asked to report tentative results. They put together what amounts to a polished rough draft, presenting their data and stating tentative conclusions and recommendations. Management then uses the information until a final report is written. The main difference between progress and interim reports is their emphasis. When a progress-report writer reaches conclusions about some aspects of the project, he or she states them but generally does not include provisional results in each report.

Initial Progress Reports

The preparation of a progress report forces you to view your project objectively and clarify your thoughts through writing. You must study the project to determine how things stand and state specifically what has been done and what needs to be done. The report speaks for the project, and if your writing lacks organization, the reader will get the impression that the project itself needs organization and direction.

Outline for Initial Progress Report

The following outline applies to initial progress reports; subsequent reports will be discussed later in this chapter. The outline can be adapted to virtually any kind of project.

```
I. INTRODUCTION

    A. PURPOSE OF REPORT

    B. PURPOSE OF PROJECT

II. WORK COMPLETED (JULY 1-JULY 31)

    A. INSTALLATION OF CUT-TO-LENGTH LINE

        Conveyor System

        Crane

    B. INSTALLATION OF BANDING LINE

        Hydraulic System

        Expansion of Banding Area

III. WORK SCHEDULED (AUGUST 1-AUGUST 31)
```

Headings under WORK SCHEDULED will be the same as for WORK COM-
PLETED, although the third-level headings (*Conveyor System,* etc.) will change
as tasks are completed and subsequent tasks are scheduled.

I. INTRODUCTION

When you write the introduction, remember that most readers are not informed about all aspects of the project. They might have walked through the
laboratory or visited the construction site, but if they really understood everything they saw, a progress report would not be necessary. The introduction
to an initial progress report requires particular care. To understand the progress
being made and the problems that will be encountered, readers must fully
grasp what the project involves.

A. Purpose of Report   A single sentence usually identifies the kind of report,
names the project, and states the time period covered by the report. If subsequent progress reports will be written, you should state the number of the
report, also.

B. Purpose of Project   Having stated the report's purpose, you must spell out
the entire project's objectives and scope; in external project reports, you often

quote directly from the contract that initiated the project. You then analyze the project, breaking it into major work areas. This analysis gives the readers a perspective of the project and prepares them for the body of the report.

II. WORK COMPLETED

The preceding outline, structured according to chronology and work areas, provides a logical organization for progress reports. Beneath the first main heading (WORK COMPLETED), second-level headings break the project into major tasks (INSTALLATION OF CUT-TO-LENGTH LINE, and so on). The scope of these major tasks should be comprehensive, allowing them to appear consistently in subsequent reports. On the other hand, third-level headings (*Conveyor System,* and so on) should be sequential, enabling you to replace them with other subtasks as the project moves forward.

III. WORK SCHEDULED

Under the heading for scheduled work, which also specifies the time period, second-level headings from the previous section appear again. They give the readers an opportunity to grasp the continuity of work in major project areas. Readers occasionally require a more detailed chronology of future work. To provide this, extend your format to include the following main headings:

```
WORK COMPLETED (DATES)

WORK SCHEDULED FOR NEXT PERIOD (DATES)

WORK PROPOSED FOR FUTURE (DATES)
```

Lengthy progress reports often have a concluding section that summarizes the overall status of the project. Instead of making a long report longer, however, consider preceding the entire report with an abstract.

Subsequent Progress Reports

Second and succeeding progress reports maintain continuity and refresh the reader's memory by adding one section, a summary of work completed prior to the present reporting period:

```
INTRODUCTION

SUMMARY OF WORK PREVIOUSLY COMPLETED (DATES)

WORK COMPLETED (DATES)

WORK SCHEDULED (DATES)
```

The new section contains the same comprehensive headings for major tasks that have appeared in prior reports but condenses the information previously presented. To prepare this section, you examine the work-completed sections of your previous progress reports and write a capsule version of them.

You also shorten your introduction in subsequent progress reports. Change only the report's number in the purpose-of-report section but reduce the purpose-of-project statement to one or two sentences. This should be adequate for the readers; they can always look in the files for the initial report's detailed description of the project.

Special Problems

Readers often indicate special interest in some aspect of the project that they particularly want to control. Budget, or any other item of special concern, can be given a main heading of its own and a thorough accounting.

A special section may also be necessary for requesting authority to change the project's scope; perhaps the investigation has produced an additional area that needs study. Before making a change that may affect the project's budget and deadline, put your recommendation in writing.

Model

JONAITIS ENGINEERING COMPANY
1715 Mandel Road, Chicago, Illinois 60646

August 1, 1978

Ms. Cecilia Roop, Chairperson
Commission for Recreation
14 Randolph Street
Chicago, IL 60601

Dear Ms. Roop:

      Subject: Progress of Feasibility Study for
                 Chicago Sports Complex (July 1–July 31)

    This is a report of progress on the feasibility study
you requested. Soldier Field, the near West Side, and the
South Loop are being studied to determine which would be best
for a sports complex. Criteria for the study are traffic accessibility and preconstruction costs, which include land
value, cost of razing, and cost of road modifications.

WORK COMPLETED

TRAFFIC ACCESSIBILITY

    Research on potential traffic patterns has been com-
pleted for one of the three alternatives. The Soldier Field
site presents severe traffic problems. The only four-lane
access road in the area is Lake Shore Drive, and traffic tie-
ups for the large complex would be even worse than ones that
occur when Soldier Field is used. No road modifications
would adequately solve the problem.
    Work on traffic patterns for the near West and South
Loop sites began September 15 and is still in preliminary
stages.

PRECONSTRUCTION COSTS

    Except for the cost of traffic modifications at the near
West and South Loop, all the cost data have been gathered and
are presented in the following table:

Ms. Roop   -2-   August 1, 1978

COST-BREAKDOWN TABLE

	Near West Side	South Loop	Soldier Field
Land	$1,000,000	$1,500,000	$1,000,000
Removal of old structures	1,000,000	1,300,000	1,800,000
Improvement of traffic facilities	TO BE GATHERED	TO BE GATHERED	2,000,000

At this point, the Soldier Field site, which would cost a to-tal of $4,800,000 and still not be adequate for reasons of traffic, has been tentatively determined unfeasible. Of the other two sites, the cost of land and razing at the near West is $800,000 less than at the South Loop, but traffic data has not been gathered.

WORK SCHEDULED (AUGUST 1–AUGUST 31)

TRAFFIC ACCESSIBILITY

Data for traffic patterns at the near West and South Loop are being gathered and processing will begin approximately October 15.

PRECONSTRUCTION COSTS

Figures for Soldier Field are complete, and cost data for the other sites are complete except for the cost of road modifications. Those data will be processed by November 1, when work on a draft of the final report will begin.

At present, this feasibility study is ahead of schedule, and a draft of the final report should be available prior to August 31.

Sincerely yours,

Robert W. Jonaitis

Writing Assignment

Your instructor may require a progress report approximately midway through your writing of a formal feasibility or proposal report. The progress report should be 350 words in length, aimed at an uninformed reader. The preceding external progress report, written in letter form, describes progress on a feasibility study. Notice the continuity between the work-completed and work-scheduled sections and the writer's use of headings to clarify the report's organization.

FRIDAY 4:00

EARLY in this text, *technical writing* was defined as writing that defies misunderstanding, and that definition applies particularly well to manuals. Companies sell not only the mechanisms they manufacture but the knowledge to use them properly. That knowledge takes the form of a manual. Both the manufacturer and the buyer want a manual that will insure safe and successful assembly, operation, maintenance, or repair of the mechanism.

In addition to training manuals, which are often indirectly related to mechanisms, the most common kinds of technical manuals, and their functions, are the following:

Assembly and Checkout: constructing, aligning, testing, and adjusting the mechanism
Operation: operating it
Service: keeping it operational through routine maintenance such as lubrication
Maintenance: locating malfunctions, testing components, and repairing or replacing them

Extremely complex mechanisms have separate manuals for assembly, checkout, operation, and so on, and sometimes several volumes for each. The most common kind of manual, however, is the operation manual. Because an operation manual must be written for virtually all mechanisms, it will be emphasized in this chapter.

The advanced mechanisms being manufactured today challenge the manual writer's ability to communicate information clearly and accurately. Manual writers obviously need to understand the mechanism thoroughly and to be aware of the readers' level of technical knowledge. Writers have no control over who will use a manual, so unless informed otherwise, they must assume that readers have limited technical backgrounds. Just as writers of formal reports must be careful with technical language, manual writers should avoid highly technical terminology, symbols, abbreviations, and mathematics whenever possible.

Thomas F. Walton, author of *Technical Manual Writing and Administration*, says the following about manual writing:

The language of technical manuals must fit the intended user. A level of writing which keeps with the average user's background, training, and intellect must be established. . . . Most technical manual programs establish a requirement that writing be aimed at a level of understanding equivalent to an eighth grade student. Therefore, the non-technical explanations and instructions must be designed so they will be understandable at this level. Also, the technical discussions must be simplified as much as possible.[1]

If some readers happen to be capable of understanding highly technical presentations, more power to them. In the manual writer's list of priorities, concern about insulting some readers' intelligence is very low.

To assist the readers, drawings and diagrams generally appear throughout a manual, providing heavy reinforcement for its descriptions and directions. While outlining a manual, the writer makes tentative decisions about visuals and then works toward integrating the drawings or photographs into the manual. Chapter 7, which included examples from a Heathkit assembly manual, provided detailed information about the effective use of illustrations.

A manual also contains headings and a rigid paragraph-numbering system to facilitate communication. After initially studying a manual, the reader uses it as a reference tool; either of the following methods, combined with a table of contents, allows rapid access to the information needed:

1.0 SECTION	1–1 SECTION
1.1 COMPONENT	1–2 COMPONENT
1.1.1 *Subpart*	1–3 *Subpart*
1.1.2 *Subpart*	1–4 *Subpart*
1.2 COMPONENT	1–5 COMPONENT
1.2.1 *Subpart*	1–6 *Subpart*
1.2.2 *Subpart*	1–7 *Subpart*

The decimal system is extremely thorough; the digit-dash-digit system depends on indentation and typographical devices to indicate parallelism and subordination.

Each major section of an operation manual builds on the previous sections. First, you introduce the manual and the mechanism. Then you must alert your reader to the mechanism's most important physical characteristics and operating principles. Finally, with the reader fully informed about the mechanism,

[1] From *Technical Manual Writing and Administration* by Thomas F. Walton. Copyright © 1968 by McGraw-Hill, Inc. Used with permission of McGraw-Hill Book Company.

you provide directions for operating it. The format of an operation manual is as follows:

Title Page
Table of Contents
List of Illustrations (when the manual contains several)
1.0 Introduction
2.0 General Description
3.0 Detailed Description
4.0 Theory of Operation
5.0 Operation

The title page, table of contents, and list of illustrations will be covered during the discussion of a student-written manual at the end of this chapter. The other five elements are explained in the following pages and exemplified by an operation manual for a compound microscope.

1.0 Introduction

The introduction's major purposes are to state the type of manual and to provide the manufacturer's name and number for the equipment. You should call the readers' attention to related manuals or publications, if any, that provide additional information about the mechanism. If appropriate, also state the level of training the readers should have to use the mechanism; for example, licenses are required to operate some electronic equipment. Finally, provide any other background or special information you may have to help familiarize the readers with the equipment.

THE COMPOUND MICROSCOPE

1.0   INTRODUCTION

This manual provides operating instructions for the Precision Instruments No. 32245 laboratory compound microscope. The microscope is used for laboratory analysis in such fields as biology, mineralogy, metallurgy, and chemistry.

2.0 General Description

In the general-description section, you actually do more defining than describing. Using the techniques you learned in Chapter 2, write a formal sentence definition of the mechanism. Then amplify, giving information that will familiarize the reader with the entire mechanism. You can provide this information in one paragraph; in fact, a manual almost invariably contains only one paragraph per heading.

The function of the general-description section is to prepare the reader for the next section, the detailed description of components. You can help accomplish this by presenting a drawing of the mechanism, often on a fold-out page that the reader can examine while studying the manual. Label the major components and give them reference numbers that you can use throughout the remainder of the manual.

2.0 GENERAL DESCRIPTION

The compound microscope is an optical instrument used in scientific procedures. Light rays pass through the specimen being examined and are altered by two lenses that create a magnified image of the object, revealing details invisible to the naked eye. Because light rays must pass through the specimen to magnify it, the compound microscope is useful only with transparent materials or very thin slices of tissue. The microscope's major components are numbered in Fig. 1.

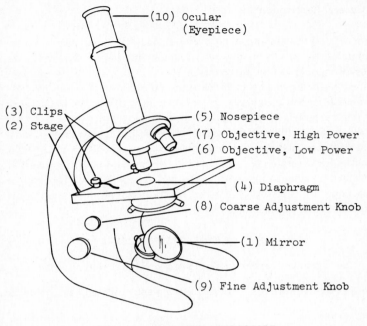

(10) Ocular
(Eyepiece)

(3) Clips

(2) Stage

(5) Nosepiece

(7) Objective, High Power

(6) Objective, Low Power

(4) Diaphragm

(8) Coarse Adjustment Knob

(1) Mirror

(9) Fine Adjustment Knob

FIG. 1:   COMPOUND MICROSCOPE

3.0 Detailed Description

The detailed-description section has many similarities with the kind of description explained in Chapter 3, but it is different in one major way. When
writing a manual, you can assume that the reader has the mechanism at hand.
Therefore, you can generally omit the description of such physical characteristics as size, weight, shape, and method of attachment of components. The
manual user, however, needs to know the function and location of major
components in order to follow the operating instructions you will provide
later.

For each major component, write a paragraph beginning with either a
formal sentence definition or an operational definition of the component.
(Operational definitions are usually sufficient for such components as adjustment knobs, controls, and gauges.) Then amplify, telling the readers everything
they need to know about each component's function and location. If you
have previously provided a drawing with labels and numbers for components,

place the reference number in the paragraph to help the reader locate the component. You can reinforce your description of a complex component by inserting a drawing right in the detailed-description section.

The order in which you describe components is very important. Select a logical order that as much as possible avoids references to components not yet described. In the microscope manual, the writer starts with the mirror, at the bottom of the mechanism, and works toward the ocular lens at the top. Had the description been from top to bottom, the writer would have needed to refer to several parts not yet described. The writer also breaks major components into subparts, providing headings for each of the subparts. This technique helps the readers by separating the information into smaller, more understandable units.

3.0 DETAILED DESCRIPTION

3.1 Mirror

The mirror (1) is the reflecting device that directs light through the object being viewed and into the micro-scope. The mirror is located at the base of the microscope and faces upward. It can be adjusted to reflect as much light as possible upward, allowing all features of the speci-men to be seen clearly.

3.2 Stage

The stage (2) is a rectangular platform on which a glass slide rests. The slide, on which a specimen is placed, is held in place on the stage by metal clips (3). The slide, with the specimen on it, is positioned over a hole in the stage known as the diaphragm (4). The diaphragm can be ad-justed to let the right amount of light (reflected by the mirror) pass through the specimen.

3.3 Nosepiece

The nosepiece (5) is a metal disk that the operator rotates to allow specific lenses to magnify an object. The
lenses located on the nosepiece are held in place by tubes
known as the objectives (6 and 7). These objectives contain
lenses of varying magnitudes. The compound microscope has a
low-power objective (6) and a high-power objective (7).

3.3.1 Low-Power Objective

The low-power objective is a lens-holding mechanism
used primarily for simple focusing. Of the two objectives, the low-power objective is the shorter and has a
magnification power of 10x. This objective is used for
basic study because its magnification is so slight.

3.3.2 High-Power Objective

The high-power objective is a lens-holding mechanism used for viewing an object in great detail. Of the
two objectives, the high-power objective is the longer
and has a magnification power of 40x.

3.4 Coarse-Adjustment Knob

The coarse-adjustment knob (8) is used to focus the microscope only when it is in low power. When turned, the
coarse-adjustment knob moves the stage rapidly up or down to
allow for focusing. Even a slight turn of the knob causes
the stage to move a great distance. As focusing begins, the

objective in use should be as close as possible to the slide
(on which the specimen rests). Then the coarse-adjustment
knob can be turned to move the stage downward. This prevents
the slide and lens from striking each other and breaking.

3.5 Fine-Adjustment Knob

The fine-adjustment knob (9) is the focusing device used
when the microscope is in high power. This knob moves the
stage only a small distance with each turn, allowing precise
focusing.

3.6 Ocular

The ocular, or eyepiece, (10) is a short tube that holds
the lens that the operator's eye directly encounters. The
lens contained in the ocular has a magnification power of
10x. This means that an operator looking through the ocular
(10x) and using the high-power objective (40x) achieves an
overall magnification of 400x.

4.0 Theory of Operation

The theory-of-operation section explains why a mechanism operates as it
does. More specifically, it defines principles that play an important role in the
mechanism's operation. The manual user does not need this information
simply to operate the mechanism; if a manual's descriptions and directions
are clearly written, an operator can usually run the mechanism and handle
emergency situations that might arise.

Knowledge of theory *does* become valuable, however, when an operator
or maintenance person follows directions for correcting a malfunction but still
cannot solve the problem. The question then becomes "What do I do now?"

and the only answer is to apply knowledge of the mechanism's principles, or the logic of its operation.

To assist the manual user, you should present theoretical information as nontechnically as possible. Rather than writing textbook-type explanations, define each principle only as it applies to the mechanism at hand. If you use a diagram to reinforce your definition of a principle, stay away from highly technical symbols. Also avoid complex mathematical equations that might confuse the reader.

Your explanation of each principle should begin with a formal sentence definition. Then elaborate, using methods of amplification covered in Chapter 2, to clarify the principle's role in the functioning of the mechanism. This will result in a paragraph for each principle, with the formal sentence definition serving as a topic sentence, as shown in the following sample:

```
                  4.0   THEORY OF OPERATION

4.1  Magnification

     Magnification is the enlargement process that increases

the apparent size of an object.  In the compound microscope,

the magnification process is performed by two lenses.  As

light rays pass through the lenses, they are refracted (bent

and spread apart) to produce a larger image.  Lenses have

different magnifications depending on how far they spread the

rays.  An image becomes inverted (upside down) as it passes

through the lenses.

4.2  Resolution

     Resolution is the separation by a lens of the light rays

of an image.  The shape of a lens determines its resolution

power.  The farther apart a lens can spread the rays, the

more sharply defined the details of an image become.
```

5.0 Operation

In a manual's operation section, you use a kind of writing rarely seen elsewhere in technical writing. It consists of directions that bluntly command the readers to perform specific operations. Grammatically speaking, you abandon the active voice, indicative mood in favor of active voice, imperative mood: "Place calibration tape on tape deck." The imperative mood implies the word *you* in front of each command, and in manuals, the articles *a, an,* and *the* are frequently omitted.

Directions for operation should be divided into short, numbered, easy-to-follow sequences as demonstrated in the microscope manual. You may sometimes tell your readers to repeat a previous series of directions, but overuse of this technique risks confusion. Within the directions, refer to components by their reference numbers to help the readers locate controls and gauges.

Your thorough knowledge of the mechanism can cause automatic writing and lead to occasional omissions. Keep in mind that if you leave out a step, the operator will probably not catch the error, and the result might be serious. Also, warn the readers about potentially dangerous operations by putting the word *warning* in capital letters and providing a short explanation of the danger.

5.0 OPERATION

5.1 Placing Slide on Stage

1. Place low-power objective (6) over diaphragm (4).

2. Turn clip (3) so it is off stage.

3. Set slide on stage with specimen centered over dia-
    phragm.

4. Return clip to original position so it holds slide
    on stage.

5. Raise stage (2) as far as possible, using coarse-ad-
    justment knob.

5.2 Focusing Microscope

WARNING

Never use the coarse—adjustment knob when microscope is in high power. The lens and slide could hit each other and break.

1. Look through ocular (10).

2. Adjust mirror (1), allowing as much light as possible to enter diaphragm.

3. Adjust diaphragm to allow sufficient light to enter and yet prevent glare.

4. Slowly lower stage, using coarse—adjustment knob, until specimen is as clear as possible.

5. Rotate nosepiece (5) so that high—power objective (7) is over slide.

6. Slowly turn fine—adjustment knob (9) until specimen is clearly in focus.

5.3 Removing Slide

1. Lower stage as far as possible, using coarse—adjustment knob.

2. Rotate nosepiece so that low—power objective is over slide.

3. Remove clip from slide.

4. Remove slide from stage.

5. Return clip to stage.

The student-written manual that follows contains a sample title page and table of contents. The manual has only one illustration, so a list of illustrations is unnecessary. If several illustrations were included, a list like the one shown would be used. This list would be placed either on a separate page after the table of contents or on the same page if doing so would not crowd the page.

LIST OF ILLUSTRATIONS

Carefully examine the manual and be prepared to discuss its strengths and weaknesses. Would an uninformed reader understand it? Is the detailed description of components organized in a way that enhances communication? If the CB unit were in front of you, would you be able to follow the instructions easily?

Model

OPERATION MANUAL

GRANADA 23-CHANNEL
CITIZENS' BAND TRANSCEIVER

by
Beth Lakin

October 20, 1978

TABLE OF CONTENTS

1.0   INTRODUCTION

This manual provides operating instructions for the Granada 23-channel citizens' band transceiver. The transceiver can be used by anyone having a class D citizens' radio service license from the Federal Communications Commission. This transceiver has 23 channels for receiving and transmitting.

2.0   GENERAL DESCRIPTION

The Granada CB transceiver (Fig. 1) is a radio used, as its name suggests, for transmitting and receiving verbal communications. It operates on a 12-volt dc power source. The CB is primarily intended for use in a vehicle with a 12-volt battery, but it can also be used in the home with the proper power attachment. The CB is compact and has a bracket for under-the-dash mounting.

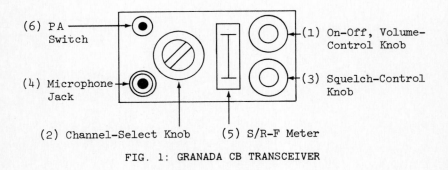

(6) P A Switch

(4) Microphone Jack

(1) On-Off, Volume-Control Knob

(3) Squelch-Control Knob

(2) Channel-Select Knob

(5) S/R-F Meter

FIG. 1: GRANADA CB TRANSCEIVER

-1-

3.0 DETAILED DESCRIPTION

3.1 ON-OFF, VOLUME-CONTROL KNOB

The on-off, volume-control knob (1), identical to the one on a regular radio, switches the transceiver's power source on or off. It also adjusts the volume of the signals received but does not adjust the volume of signals transmitted.

3.2 CHANNEL-SELECT KNOB

The channel-select knob (2) rotates so the operator can select the channel on which to transmit and receive. When the operator places the selector on 3, for example, the unit will receive stations on that channel and the operator can transmit on that channel.

3.3 SQUELCH-CONTROL KNOB

The squelch-control knob (3) is used to control static and background noise from the receiver. When properly adjusted by its operator, the squelch automatically performs two functions: it switches off the sound to the loudspeaker when no stations are on the channel, and when someone calls, it senses the signal and opens the speaker.

3.4 MICROPHONE

The microphone converts verbal signals into electrical impulses for transmission. The microphone operates with a "push-to-talk" button. To transmit, the operator holds the

button down. Releasing the button switches the unit to "receive" so incoming signals can be heard. The microphone
plugs into a microphone jack (4) so it can be removed to prevent theft.

3.5 S/R-F METER

The S/R-F meter (5) shows the input-signal strength (S)
when receiving and the range-frequency (R-F) output when
transmitting. The meter is marked in units from 1 through 9,
and a low reading indicates that the system is not receiving
or transmitting at normal range.

3.6 PA SWITCH

The PA (public-address) switch (6) can convert the CB
transceiver into a miniature public-address amplifier. An
extra 8-ohm speaker is needed to complete the PA system, however; a PA speaker jack is provided on the back panel of the
CB for connection to such a speaker. The PA button must be
pushed to amplify the operator's voice.

4.0 THEORY OF OPERATION

4.1 TRANSMISSION

Transmission with a CB unit is the process of converting
verbal signals into electrical impulses that travel from one
CB antenna to others on the same channel (frequency). Verbal
signals are converted into electrical impulses by the operator's microphone. The electrical signals are then strength-

ened within the CB unit and sent out through the antenna.

4.2 RECEPTION

Reception with a CB unit is the process of converting electrical impulses into audible sounds. The CB antenna picks up incoming electrical signals. Within the transceiver, they are converted into verbal signals, and the sounds are heard through the CB speaker.

5.0 OPERATION

5.1 TO RECEIVE

1. Turn unit on with on-off, volume-control knob (1).
2. Rotate squelch-control knob (3) fully counterclockwise.
3. Set channel-select knob (2) to desired channel.
4. Advance volume-control knob (1) until background noise is heard from speaker.
5. Adjust squelch-control knob (3) clockwise until receiver's background noise is muted (squelched).

5.2 TO TRANSMIT

1. Select channel with channel-select knob (2).
2. Wait until channel is clear.
3. When channel is clear, depress microphone button and speak in normal tone of voice with microphone held 2-3 inches from mouth.
4. To receive, release microphone button.

Writing Assignment

Select a mechanism you frequently use when working on projects in your technical field. Write a 1000-word manual aimed at a high school sophomore who has limited technical knowledge and has not previously operated the mechanism. Assume that the sophomore is your assistant and that the success of your next project depends on his or her skill.

Provide a title page and table of contents, and include the five sections emphasized in this chapter. The operating instructions should be for procedures commonly performed with the mechanism.

Exercise

Write a step-by-step set of directions for changing a flat tire, aiming your writing at someone who has never done it. To make the directions easy to follow, break them into logical sequences such as (1) assembling jack, (2) positioning jack, (3) removing flat tire, (4) mounting new tire, (5) and disassembling jack. Your set of directions, which should total approximately 50 steps, may include the following two drawings.

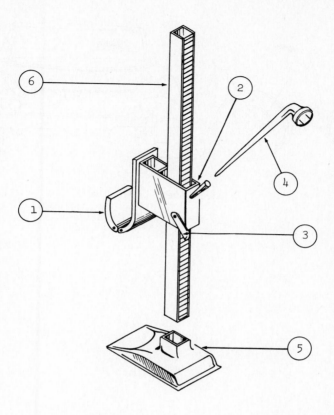

(1) Jack Hook   (4) Jack Handle

(2) Jack Head   (5) Jack Base

(3) Jack Locking Lever   (6) Jack Post

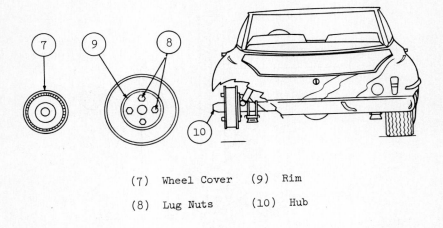

(7) Wheel Cover   (9) Rim

(8) Lug Nuts   (10) Hub

14   Oral Reports

AS you advance in industry, your ability to speak convincingly, as well as your writing skill, becomes increasingly important. At conferences, you are called on to explain the results of investigations, propose solutions to problems, report on the progress of projects, and justify your department's requests for more employees and equipment. Every kind of formal, written report has its verbal counterpart; sometimes an oral report supplements a written one, and often a verbal presentation takes the place of written reports. Although you use the same information-gathering process for both oral and written communication, you must organize the information differently.

Oral and Written Reports

The most significant difference between oral and written communication is also the most obvious: an oral report has listeners, and a written one has readers. But why not use the same report for both kinds of presentations? How is a listener different from a reader?

1. A listener is present for the entire oral report. This would seem advantageous for a speaker, but it actually makes verbal communication more difficult. Let's assume that your listeners are the plant manager and his or her staff. It is unlikely that the plant manager wants to hear your entire report. He or she would probably prefer a capsule version of the report, and a written report's abstract would provide it. The manager would then order the staff members to examine the details contained in the body of the report. An oral report, however, gives the manager no choice but to listen to all your information.

The manager's staff members are not interested in the entire report, either. If the report were written, each of them would read the abstract and use the table of contents to locate the financial, technical, personnel, or other section of the report that interests her or him. All the staff members are present for the entire oral report.

2. Your listeners have only one opportunity to grasp the information. Even though a question-and-answer period may follow the report, listeners cannot study the information as a reader would study a formal report.

3. An oral report does not provide headings to identify sections of particular interest to the listeners and to indicate parallel and subordinate ideas.

These differences emphasize the problems that exist in speaker-listener situations. Speakers, however, also enjoy many advantages. They can use their personality, voice, and gestures, as well as first-person pronouns, visuals, and feedback from listeners. The remainder of this chapter explains how you can effectively organize and communicate your information in an oral report.

The Extemporaneous Report

Many inexperienced speakers want to write every word of their reports. They are afraid of forgetting important ideas and think the solution is either to read or to memorize the report. However, most experienced speakers agree that the extemporaneous report, which is outlined rather than read or memorized, is the most effective kind of presentation. The extemporaneous method emphasizes natural, conversational delivery and concentration on the audience. Using this method, you can direct your attention to the listeners, referring to the outline only to jog your memory, and insure that ideas are presented in the proper order. Your word choice occasionally suffers, but words spoken extemporaneously still communicate better than words memorized or read.

The best way to combat fear and forgetfulness during an extemporaneous report is simply to know your subject. If you are the best-informed person in the room, you can stop worrying about your subject and start concentrating on communicating. The outline for an extemporaneous speech contains main headings and subheadings; you can develop the headings into sentences during rehearsals.

INTRODUCTION

In industry, do not worry about finding a humorous story, a quotation by an authority, or a recent occurrence to begin your report. At the outset, you already have your listeners' attention. Capitalize on that by saying something strongly related to your topic. If your opening comments are interesting or amusing but not relevant to the report, you have to add a second attention step to transfer the listeners' attention to your subject. A good opening may simply identify your subject and purpose, show how the subject affects the listeners, or explain why the investigation was undertaken.

The introduction must also give your listeners a clear idea of the order in which you intend to present your ideas. For a recommendation report, this requires stating the possible alternatives and specifying your criteria for judging them. You can introduce a proposal by identifying the major areas that you will discuss. In any oral report, the listeners must clearly understand your plan of development so that they can grasp the information in the body of the report.

BODY

Time is to an oral report what space is to a written one. An oral report, however, does not lend itself to the concise type of presentation you use in a written report. To communicate an idea verbally, you must state a generalization, provide details to support it, and reinforce it with a summary. Numerous studies have shown that listeners simply do not hear everything you say, and if they miss an idea or an important detail, they have no recourse. Therefore, the verbal communication process usually consumes several minutes per main idea.

Always impose your own time limit on the report, and narrow your number of main ideas accordingly. It is much better to present two or three main ideas carefully than to attempt to communicate more information than your listeners can grasp. If you select only the most important ideas, your speech will be limited enough to please the plant manager and detailed enough to satisfy his or her staff members. Throughout the report, avoid potentially confusing terms, and if possible, reinforce your main ideas with visuals.

CONCLUSIONS

As you conclude your report, you should actually say "In conclusion . . ." in order to capture your listeners' interest. The concluding section emphasizes the main ideas presented in the body of the report. If your objective is persuasion, stress the ideas' main advantages and urge your listeners to take specific action. For a recommendation report, emphasize the most significant data presented for each criterion and clearly present your recommendations. You can often use a visual to summarize the important data presented in either a proposal or a recommendation report. End the report by asking if your listeners have any questions.

Visuals

Visuals are invaluable for emphasizing and clarifying complex ideas in an oral report. As you construct your outline, determine which ideas will benefit from visuals and build the report with them in mind. The visuals should be aimed at the technical level of your listeners; as Chapter 7 explained, the same information can be presented in either simple or complex form. Also, use several simple visuals instead of crowding too much information on one, and place as few words as possible on the visuals—you are responsible for verbal interpretation. Color adds life to visuals, but black and white offers clarity if some listeners are far away.

The size of the audience partially determines the method of showing or projecting visuals. Flip-charts are easy to manipulate and effective for small groups. For small or large groups, consider slide, opaque, or overhead projectors. You can regulate the size of the projected images by simply moving the projector toward or away from the screen and adjusting the focus. The room must be semidark for slide and opaque projections. Overhead projections can be used in normally lighted rooms, however; nobody has to flip light switches, and the audience is not distracted. Overhead projectors require transparencies, but making them is a simple process, and they are easy to handle during the presentation.

Rehearsals

To be successful, extemporaneous reports must be thoroughly rehearsed. During rehearsals, go straight through the speech, using small note cards. In each trial run, attempt to give the presentation a conversational quality, and practice using your voice and gestures to emphasize important points. You have rehearsed enough when you feel secure with your report, but always stop short of memorization; if you do not, you will ultimately grope for memorized words rather than concentrate on the listeners and let the words flow.

Final rehearsals for reports to large groups should simulate conditions under which the speech will be made. This includes a room of approximately the same size with the same type of equipment for projecting your voice and visuals. Rehearsals of this type not only guard against technical problems but allow you to become comfortable in an environment similar to the one for the final report. Ask a colleague to attend at least one rehearsal, comment on how well he or she can hear you and see the visuals, and offer a critique of the speech, including any possibly distracting mannerisms.

Delivery

A report can be no better than its preparation, but the following is a list of suggestions to keep in mind as you face your listeners and deliver the speech.

1. Do not start the speech with an apology. An audience is usually more willing to accept your information if you speak positively.
2. Make sure you can be heard, but try to speak conversationally. The listeners should get the impression that you are just talking to them rather than that you are making a report. Inexperienced speakers often talk too rapidly.
3. Look directly at each listener at least once during the report. With experience, you will be able to tell by the viewers' faces whether you are communicating.
4. Fight the tendency to use your outline when you do not need it. When collecting your thoughts, do not say "uh"; pause and remain silent.
5. When finished with a visual, remove it so that it does not compete with you. If you are using a pointer, set it down to avoid tapping with it.
6. If you cannot answer a question during the question-and-answer session, say so and assure the questioner that you will find the answer.

Speaking Assignment

Your instructor may require an oral presentation of a formal report you have written during the term. The speech should be extemporaneous and approximately 10 minutes long. To prepare your presentation, follow the suggestions in this chapter for converting a written report into a successful oral presentation. Make your visuals, outline the speech, and most important, rehearse. A question-and-answer session with other members of the class should follow your presentation.

Section Five
Letters

In this section, the text moves away from essentially data-oriented reports to people-oriented letters. First, the various letter formats and the elements of a letter are described. Then, the kinds of business letters are explained, with emphasis on the indirect approach for letters of persuasion, and the direct approach for informative letters. Application letters, perhaps the most important letters you will write, are emphasized in the final chapter.

JFC *JONAITIS E*

1715 Mandel Road, Chicago, Ill

Heading

↑

4-8 spaces

↓

Ms. Cecilia Roop, Chairperson
Commission for Recreation
14 Randolph Street
Chicago, IL 60601

Dear Ms. Roop:

15   Letter Formats

A LETTER'S appearance gives the reader a first impression of the writer and the firm he or she represents. Contributing to the visual attractiveness of a letter are its letterhead, balance, neatness, and format. The format must also be appropriate to the letter's contents. This chapter begins by discussing the common formats for business letters: the block, full-block, modified-block, and simplified formats. It then describes and exemplifies the elements that comprise a letter's format.

The Four Formats

In the block, full-block, and modified-block formats, the word *block* refers to the shape of the letter's major sections. When the left side of a section has no indentation, the section is considered block shaped.

BLOCK

The *block format's* inside address, body, and signature sections have no indentation (Fig. 15.1). The block for the inside address and body are flush with the left margin, and the left side of the signature block begins at the center of the page. As with the full- and modified-block formats, if the letter contains no letterhead, its heading includes the writer's address as well as the date, and they form a block that ends at approximately the right margin.

FULL BLOCK

In the *full-block format* (Fig. 15.2), each element of the letter is flush against the left margin. Thus, the left side of the letter forms a full, or complete, block. The full-block format lacks balance but is probably the most widely used because it can be quickly typed.

MODIFIED BLOCK

Placement of the blocks in the *modified-block format* (Fig. 15.3) is identical to that of the block format, which causes some confusion between the two. The first line of each paragraph in a modified-block format is indented, modifying the block. This format is also called the semiblock format.

ⒿⒺⒸ *JONAITIS ENGINEERING COMPANY*

1715 Mandel Road, Chicago, Illinois 60646

↓
4-8 spaces
↑

↓ [Heading] ───────→ August 29, 1978
4-8 spaces
↑

Ms. Cecilia Roop, Chairperson
Commission for Recreation ←────── [Inside
14 Randolph Street   Address]
Chicago, IL 60601
2 sp.
Dear Ms. Roop: ←────────── [Salutation]
2 sp.
Subject: Feasibility Report on Sports Complex ←─ [Subject Line]
2 sp.
As you requested on June 4, 1978, I am submitting this feasibility report, entitled "Site for a Chicago Sports Complex."

The report examines the feasibility of three proposed sites:
Soldier Field, the near West Side, and the South Loop. As
you directed, emphasis has been placed on traffic accessibility and preconstruction costs, and the recommendations are
based on those criteria. New city expressways are presently
being proposed, but they remain an uncertain variable and are
not considered in this study.

I am available for consultation about this report and for
further research as you move forward with the sports complex
project.

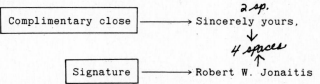

[Complimentary close] ─────→ *2 sp.*
Sincerely yours,
↓
4 spaces
↑
[Signature] ─────→ Robert W. Jonaitis

Fig. 15.1   Block format

JONAITIS ENGINEERING COMPANY
1715 Mandel Road, Chicago, Illinois 60646

August 29, 1978

Ms. Cecilia Roop, Chairperson
Commission for Recreation
14 Randolph Street
Chicago, IL 60601

Dear Ms. Roop:

Subject: Feasibility Report on Sports Complex

As you requested on June 4, 1978, I am submitting this feasibility report, entitled "Site for a Chicago Sports Complex."

The report examines . . . .
. . . .

Sincerely yours,

Robert W. Jonaitis

Robert W. Jonaitis

Fig. 15.2   Full-block format

JONAITIS ENGINEERING COMPANY
1715 Mandel Road, Chicago, Illinois 60646

August 29, 1978

Ms. Cecilia Roop, Chairperson
Commission for Recreation
14 Randolph Street
Chicago, IL 60601

Dear Ms. Roop:

    Subject: Feasibility Report on Sports Complex

    As you requested on June 4, 1978, I am submitting this feasibility report, entitled "Site for a Chicago Sports Complex."

    The report examines . . . .
. . . .

                Sincerely yours,

                Robert W. Jonaitis

Fig. 15.3 Modified-block format

SIMPLIFIED

The *simplified format* (Fig. 15.4) is a recent development that departs from conventional letter formats. Its streamlined format contains no salutation and no complimentary close, but it almost always includes a subject line. Advocates of this format consider it more efficient than the other three because it requires less typing. A letter's efficiency, however, should be determined by its results rather than its typing time. The success of a simplified letter depends on the writer's ability to overcome the format's extremely direct approach and establish rapport with the reader. Rapport is not crucial in letters containing routine or positive information, but the simplified format would seem to present problems in letters requiring persuasion. Such letters need an indirect approach, and there is nothing indirect about a subject line. Direct and indirect approaches will be discussed more specifically in Chapter 16.

Elements of a Letter

As Fig. 15.1 shows, letters of any length can be balanced fairly well by adjusting the space between the letterhead and the date, and between the date and the inside address. You can also achieve balance by widening or narrowing the letter's margins. Unless the letter is extremely short, you should single-space the body, with two spaces between paragraphs.

HEADING

Because virtually all companies use letterhead stationery, the heading of a business letter contains only the date, which ends at about the right margin. In personal letters, include both the address and the date:

```
4217 East Avenue
Claremont, In 46327
May 15, 1978
```

A complete personal letter appears at the end of Chapter 17.

INSIDE ADDRESS

Readers are sensitive about their names, titles, and firms, so the inside address requires special care. Make sure that you use the correct personal title (Mr., Ms., Dr., Professor) and business title (Director, Manager, Treasurer). Write

JONAITIS ENGINEERING COMPANY
1715 Mandel Road, Chicago, Illinois 60646

August 29, 1978

Ms. Cecilia Roop, Chairperson
Commission for Recreation
14 Randolph Street
Chicago, IL 60601

Subject: Feasibility Report on Sports Complex

As you requested on June 4, 1978, I am submitting this feasibility report, entitled "Site for a Chicago Sports Complex."'

The report examines the feasibility of three proposed sites:
Soldier Field, the near West Side, and the South Loop.   As
you directed, emphasis has been placed on traffic accessibility and preconstruction costs, and the recommendations are
based on those criteria.   New city expressways are presently
being proposed, but they remain an uncertain variable and are
not considered in this study.

I am available for consultation about this report and for
further research as you move forward with the sports complex
project.

Robert W. Jonaitis

Robert W. Jonaitis

Fig. 15.4   Simplified format

the firm's name exactly, adhering to its practice of abbreviating or spelling out such words as *Company* and *Corporation*.

The reader's business title can be placed after his or her name, on a line by itself, or preceding the name of the firm, whichever best balances the inside address. For example:

```
Ms. Marian Collins
Director of Planning
American Rail Company
9115 Charles Avenue
Lee, CA 93541
```

The U.S. Postal Service recommends that two-letter abbreviations be used for each state. These abbreviations should be used on envelopes, in headings, and in inside addresses. As shown in the following list, both letters of the abbreviation should be capitalized; two spaces should separate an abbreviation from a zip-code number.

Alabama	AL	Michigan	MI
Alaska	AK	Minnesota	MN
Arizona	AZ	Mississippi	MS
Arkansas	AR	Missouri	MO
California	CA	Montana	MT
Colorado	CO	Nebraska	NE
Connecticut	CT	Nevada	NV
Delaware	DE	New Hampshire	NH
District of Columbia	DC	New Jersey	NJ
Florida	FL	New Mexico	NM
Georgia	GA	New York	NY
Guam	GU	North Carolina	NC
Hawaii	HI	North Dakota	ND
Idaho	ID	Ohio	OH
Illinois	IL	Oklahoma	OK
Indiana	IN	Oregon	OR
Iowa	IA	Pennsylvania	PA
Kansas	KS	Puerto Rico	PR
Kentucky	KY	Rhode Island	RI
Louisiana	LA	South Carolina	SC
Maine	ME	South Dakota	SD
Maryland	MD	Tennessee	TN
Massachusetts	MA	Texas	TX

Utah	UT	Washington	WA
Vermont	VT	West Virginia	WV
Virginia	VA	Wisconsin	WI
Virgin Islands	VI	Wyoming	WY

ATTENTION LINE

Attention lines, which appear two spaces below the inside address, are generally used only when you cannot name the reader ("Attention Personnel Manager"; "Attention Payroll Department"). The word *Attention* is usually placed against the left margin and is not followed by a colon.

SALUTATION

The salutation always agrees with the first line of the inside address. If the first line names an individual (Ms. Marian Collins), say "Dear Ms. Collins." When the first line names a company, the trend is toward repeating the name of the company ("Dear American Rail Company" or "American Rail Company") instead of saying "Gentlemen," which assumes that anyone important at the company is male. A colon always follows the salutation.

```
American Rail Company
9115 Charles Avenue
Lee, CA 93541

Attention Ms. Marian Collins

American Rail Company:
```

SUBJECT LINE

Subject lines, which allow a direct approach to the subject of the letter, are becoming common in business letters, particularly information letters. Place the subject line ("Subject: Training Program") two spaces below the salutation, and indent it if you indent the letter's paragraphs. Subject lines are often either underlined or completely capitalized.

COMPLIMENTARY CLOSE AND SIGNATURE

Use simple closings, such as "Sincerely yours" or "Yours truly," to end business letters. Capitalize only the first word, and place a comma after the closing.

Company policy generally specifies what to place in the signature section. It sometimes requires that the company's name be placed immediately below the complimentary close. The writer's title or department, or both, often appear below her or his typed name.

Yours sincerely,
DAVIS MANUFACTURING CO.

Ronald C. Anderson

Ronald C. Anderson
Personnel Director

Yours truly,

William J. Brockington

William J. Brockington
Manager, Drafting Division

SUCCEEDING PAGES

For succeeding pages of a letter, place the name of the reader, the page number, and the date in a heading:

Ms. Collins   —2—   May 15, 1978

AWARENESS of the reader, which has been emphasized in each of the report-writing chapters, is even more important in writing business letters. Although both letters and reports require anticipation of the reader's needs, letters are generally more people oriented: the letter reader's reaction to the *way* you present information is often more important than the information itself. In letters and in everyday conversation, how many times have you been disturbed not so much by what someone has said, but by the way it was said? This chapter explains how to word and organize your letters so they will achieve their purpose and avoid alienating the reader.

Style

In business letters, use the first-person *I* and *we*, and refer to the reader in the second-person *you*. This is the first step toward giving your letters the natural, conversational style needed for personal communication. The second step is to write in plain English. The "businessese" style of writing convinces your reader that you regard him or her as a number rather than as a person:

Pursuant to our discussion of February 3 in reference to the L–19 transistor, please be advised that we are not presently in receipt of the above-mentioned item. Enclosed herewith please find a brochure regarding said transistor as per your request.

Such stilted, pompous prose is so common in business correspondence that many young writers think they are supposed to write that way. Why not say, "I've enclosed a brochure on the L-19 transistor we talked about on February 3. Our shipment of L-19's should arrive within a week." Recall the last bit of correspondence you received from your school; did it sound like a human being had written it? When you write letters in business and industry, remember that a conversational style invariably achieves better results than businessese.

Tone

Tone refers to your attitude toward the subject and reader, as reflected in your choice of words and your overall approach to the letter-writing situation.

When you write a letter, you are the message; you must demonstrate your understanding of the reader's point of view by using the "you" approach, and you must also use a positive approach to reveal your company's point of view.

THE "YOU" APPROACH

On one level, the you approach requires only a common-sense awareness of human nature. No reader wants to be treated like a number, taken for granted, told how to do a job, or threatened. On the other hand, human nature occasionally tempts all of us to do those very things, particularly in frustrating letter-writing situations. When that happens, remember that your letter represents not only yourself but your firm. The "I-guess-I-told-you" letter that sometimes makes you feel good generally does not get the results your company wants. The reader will be alienated unless you show common courtesy and respect.

On another level, the you approach requires a thorough analysis of each reader's point of view, which is invariably different from that of your firm. For example, let's assume that your firm has sold a conveyor system to Calvin Number. To your firm, Number represents just another conveyor system, a very small item on your profit-and-loss ledger. For Number's company, of course, the conveyor might represent last month's profits. If the conveyor is defective, it becomes one of the many routine problems that your firm handles every day. The defective conveyor might not be routine for Number, however; perhaps he is very worried about it. To maintain a good business relationship with him, your letter must show your understanding of his position. The defect matters to him, and he wants to know that it has some importance for you, also. By putting yourself in Calvin Number's position, you will be able to anticipate his reaction to your letter and to insure that you do not evoke a negative response. A you-approach letter requires time and thought, but it helps establish and maintain good business relationships.

THE POSITIVE APPROACH

Industry has no room for negative thinking, and it has even less room for negative letter writing. From your firm's point of view, letters to customers and other firms are part of public relations; ultimately, they can affect profits. Therefore, all business letters are sales letters that promote or protect the image of your firm. To this extent, they are similar to external proposals, which attempt to win contracts by communicating a "we-can-handle-it" attitude. In letters that must convey negative information about your firm or one of its customers, never misinform the reader but present the information in a positive rather than negative light. Avoid the following phrases in business letters because they have negative connotations:

> This problem . . . (admits that there *is* a problem)
> You won't be sorry . . . (suggests that the reader *might* be sorry)
> Your failure . . . (implies that the reader is a failure)
> I doubt . . . (conveys negative attitude; say "I believe . . .")
> You claim . . . (implies that the reader has lied)

If you are not convinced of industry's emphasis on the positive approach, try beginning an application letter with "I'm afraid I haven't had much actual working experience . . .," and see what happens, or more specifically, see what does not happen.

Direct or Indirect Structure?

The main function of every business letter is either to inform or to persuade. To determine the main function of your letter, decide whether the reader will react positively or negatively to the information it contains. If the reader will react negatively, give the letter an *indirect structure*: spend at least a paragraph preparing him or her for the information or persuading him or her to accept the information. On the other hand, if the reader will respond positively, give the letter a *direct structure*: present the information immediately. The form, or structure, of a letter is always determined by its main function; form follows function.

The rest of this chapter explains the functions of the most common types of business letters and suggests how to structure their contents.

LETTERS OF INQUIRY

Letters of inquiry from your company to another should be given a direct structure. Usually, you are asking about the other firm's product or service, and you do not have to convince the reader to provide the information; firms view letters of inquiry as opportunities to gain customers. Therefore, simply identify the information you need in a one- or two-sentence letter.

As a student, you may occasionally write a letter of inquiry requesting that a company send you information for use in a class project or report. Use the direct approach because large firms receive many requests of this type and appreciate a courteous but concise letter. State specifically the information you need, why you need it, and why you selected the particular firm as a source of information. In closing the letter, you may say that you look forward to hearing from the firm, but do not thank them in advance. Also, do not enclose a stamped, self-addressed envelope unless you are writing to a small firm. Figure 16.1 is a letter of inquiry from a company requesting special information from another company.

ANSWERS TO INQUIRIES

Except for a letter responding to a student's request, an answer to an inquiry is usually a sales letter. Unlike most sales letters, however, it does not require an indirect approach; having solicited information, your reader will be receptive to it. Answer an inquiry immediately, if only to acknowledge the letter and explain that you have referred it to another department or dealer. An inquiry gives you some insight into the reader's needs, so inject the you approach immediately after thanking him or her for writing. As Fig. 16.2 shows, an answer to an inquiry should describe your product in terms of the reader's advantage, emphasizing its strong points. Enclose a brochure or other sales literature with your letter, and insure that the reader will have no difficulty ordering the product.

TRANSMITTAL LETTERS

A transmittal letter conveys a report from one firm to another. It is an information letter that has a direct structure. Begin the letter by identifying the enclosed report and stating the date it was requested. Then, provide a brief paragraph or two explaining the report's purpose and scope. Close the letter by indicating your availability if the reader has any questions about the report. (Chapter 9 described transmittal correspondence in detail and Chapter 15 contained an example of a transmittal letter.)

CREDIT LETTERS

A credit letter responds to a request for credit from a firm or individual. The structure of the letter depends on whether your response is positive or negative. A letter *granting* credit should be given a direct approach; the reader will obviously be receptive to your information. Welcome the reader as a

Larson
Electronics
Corp.

190 Jackson Blvd.
Atlanta, Ga.   30307
404-627-6335

April 5, 1978

Mr. Lawrence M. Callison
Personnel Director
Carlton Bridge Company
9311 Commerce Avenue
Boston, MA 02107

Dear Mr. Callison:

May I ask a favor of you? James Creekmore, a representative of the Melton Corporation, recently told me of your success in managing Carlton's training program for the hardcore unemployed. Our firm is presently planning such a program, and we would appreciate information about the attitude-changing section of your program.

We expect to begin our training sessions on approximately June 1. I believe we are prepared for the remedial-education and job-skills portion of the program, but we need help with attitude-changing. Specifically, we would appreciate knowing who you employed to teach that section, the major emphasis in the section, and the length of the section in relation to the entire program.

I congratulate you on your success in managing Carlton's program, and I believe your ideas would help us solve this long-neglected problem, also. I look forward to hearing from you.

Sincerely yours,

Jane F. Weathers

Jane F. Weathers
Project Coordinator

Fig. 16.1   Letter of inquiry

Andersen ◆A/W.◆ Windowalls®

ANDERSEN CORPORATION
BAYPORT, MINNESOTA 55003 · PHONE 439-5150 · CODE 612

April 4, 1978

Mr. Jerry Mikulski
JM Company
5524 Claude Ave.
Hammond, IN 46320

Dear Mr. Mikulski:

Thank you for writing to us requesting information concerning
Andersen Products.

Enclosed you will find our catalog with complete information
regarding features, sizes and details of all Andersen windows, gliding doors, and shutters.

All styles and sizes of window and gliding door units we manufacture are illustrated in the enclosed information. We do
not assemble custom windows or gliding doors to fit specified
or existing openings.

Our products are shipped to our distributors located throughout the United States. Our distributors, in turn, service
various lumber and millwork dealers in their area. It is for
this reason that we do not maintain a consumer retail price
list. Price information for our products can be obtained
through your local dealer. If the dealer cannot answer all
your questions, we suggest you contact one of the distributors listed on the enclosed sheet.

We appreciate your interest and consideration of Andersen,
Mr. Mikulski, and if we can be of further help, would you
please let us know.

Yours truly,

ANDERSEN CORPORATION

Rick Hoffbeck

Sales Promotion Department

Rick Hoffbeck: SPD

Fig. 16.2 Answer to an inquiry. Courtesy Andersen Corporation, Bayport, Minnesota.

credit customer and explain the terms of credit. When *refusing* credit, however, an indirect structure is necessary. Thank him or her for requesting credit, and avoid insulting phrases like "bad risk" or "inability to pay debts promptly." In the second or third paragraph of your letter, inform the reader that you are unable to grant the request, but use the positive approach. As Fig. 16.3 shows, you should also encourage the reader to buy on a cash basis until he or she has built up a credit rating, and express the hope that you will soon be able to extend credit.

ADJUSTMENT LETTERS

An adjustment letter responds to a customer's claim that your firm owes him or her something. The customer generally believes that you have sold a defective product and wants you to replace it, repair it, or return the money. If your investigation shows that the claim is justified, thank the customer for calling it to your attention, apologize for the mistake, and state the terms of the adjustment. Never guarantee that the mistake will not happen again unless you are prepared to honor the guarantee.

A letter refusing adjustment must have an indirect structure, as shown in Fig. 16.4. Thank the reader for writing, express your understanding of the situation, and explain what action you have taken on her or his behalf. Gradually build up to your refusal by explaining why the adjustment cannot be granted. Close the letter by stating that your firm appreciates the customer's patronage and looks forward to future business with him or her.

SALES LETTERS

Except in response to an inquiry, a sales letter must have an indirect structure. Readers receive hundreds of sales letters and must be enticed into reading them. A sales letter contains the following elements:

Attention: arouses the reader's interest in the product
Desire: uses the you approach by describing the product in terms of the reader's advantage
Conviction: provides details to convince the reader that the product is the best of its kind
Action: urges the reader to buy the product and makes it easy to do so

November 14, 1978

Mr. Mark L. Smith
7635 Sloan Ave.
Hammond, IN 46324

Dear Mr. Smith:

Thank you for your patience while we have been making our
credit check for your account.

We have gathered whatever information we could find to serve
as a basis for credit. However, we do not feel that we can
open an account for you at this time due to the lack of
credit references.

If you have information that has been unavailable to us, we
would be very glad to receive it as soon as possible. In the
meantime, we invite you to continue as a customer of ours on
a cash basis.

We look forward to serving you in the near future.

Sincerely yours,

Connie L. Chen
New Accounts Supervisor

Fig. 16.3 Letter refusing credit

JONAITIS ENGINEERING COMPANY
1715 Mandel Road, Chicago, Illinois 60646

August 9, 1978

Mr. Daniel Redling
General Manager
Klein Leather Company
311 Miller Ave.
Morris, IN 46325

Dear Mr. Redling:

Thank you for your letter of August 6. I appreciate the opportunity to explain the $745 bill for constructing the railing in your plant.

You are correct in stating that the price quoted to you over the phone was $600. On examining the site, however, our project engineer informed your office that stronger material would be required to insure the safety of the railing. With the approval of your office, an additional $145 went into the stronger material.

Again, I am pleased to be able to clear up this misunderstanding. Jonaitis Engineering looks forward to working for you in the future.

Yours sincerely,

Eric I. Cleese

Eric I. Cleese

Fig. 16.4   Letter refusing adjustment

Dear Reader:

   You wouldn't consider buying a house without going through it from top to bottom to make sure it offers exactly what you're looking for.

   The same rule holds true for a book: Look before you buy.

   That's why I'm always surprised when people don't take us up on our 15-day free-trial invitation.

   Surely if you heard about the house of your dreams at a ridiculously low price, you wouldn't pass up the chance to look at it. So how can you refuse an invitation to examine a series that could save you hundreds, even thousands of dollars—could <u>give</u> you the house of your dreams—without risking a single dime?

   If you think no book can teach you to do major repairs, think again. <u>Space and Storage</u>, along with subsequent volumes in the HOME REPAIR AND IMPROVEMENT series, will guide you through jobs you might never have dreamed of tackling before —with step-by-step instruction as easy to follow as anything that has ever appeared in print.
   But the point is: Don't take my word for it. Look before you buy. And if you don't like what you see, just slip the book back into its carton and return it to us.

   It's a practical offer for practical people . . . and we wouldn't dare to make it if we couldn't back it up.

   Just initial the card and drop it into the mail.

                    Sincerely,

                    Joan D. Manley
                    Joan D. Manley
                    Publisher

JDM/ASC

Fig. 16.5   Sales Letter. Courtesy Time-Life Books, Inc.

Apply the you approach throughout the letter but particularly in the attention and desire steps. In firm-to-firm sales letters, factual appeals to the reader's desire for profit, economy, and efficiency are the most effective. Firm-to-individual letters, in addition to emphasizing economy, generally appeal to the reader's desire for such things as prestige, sexual attractiveness, and comfort. A sales letter is shown in Fig. 16.5.

COLLECTION LETTERS

Depending on how a company views its customers, it uses one of two methods for collecting money. It either makes a strong effort to collect the money immediately, or it attempts to keep the customer *and* collect the money. The direct approach simply threatens the reader with either a suit or a collection agency. The indirect approach tries to avoid demanding the money. A firm that uses this approach wants the reader to do three things: keep the product, pay for it, and buy more products from the firm. The company's first collection letter assumes that the reader has forgotten the payment or wants to be reminded of it. The next letter assumes that the customer has not paid because the product is defective. Finally, as a last resort, the firm threatens the reader. When the indirect approach works, it generally accomplishes the firm's three goals. On the other hand, a direct letter often loses the customer even when it succeeds.

Writing Assignment

1. You are a research analyst at Gabriel Corporation, which is building a carbon monoxide plant near Boise, Idaho. The Occupational Safety and Health Act of 1970 requires that the new plant have a safety program for checking employees' exposure to carbon monoxide. At the Environmental Research Center in Cleveland, Ohio, Dr. Marlon T. Jenner has been conducting tests on the effects of carbon monoxide. The results of his studies have been published in several periodicals. Write a letter to Dr. Jenner requesting that he assist you in setting up the safety program. Gabriel Corporation is prepared to pay Jenner's fee for consulting.

Dr. Marlon T. Jenner
Director of Research
Environmental Research Center
3300 West 19th St.
Cleveland, OH 44109

2. Mel Grant, a lift-truck operator for Metro Compressed Gas Company, has been fired for working while under the influence of alcohol. Because Grant was extremely popular with the other employees, his dismissal has damaged their morale. Unfounded rumors that other firings are imminent have made the problem worse. In your position as department supervisor, you first considered using the regular company bulletin to inform employees of the facts. You rejected this idea, however, because the bulletin would probably be viewed as company propaganda.

Write a letter to William Swikert, the informal leader of the employees, explaining Grant's firing to him and requesting that he present the company's view to the workers.

Mr. William Swikert
2367 Livingston Ave.
Chicago, IL 60618

17 Letters of Application

A SUCCESSFUL letter of application and data sheet must be preceded by self-analysis and by investigation of prospective firms. Both are prerequisites to the you approach explained in Chapter 16. *Self-analysis* means adopting the objectivity of prospective employers and determining what you have that will appeal to their self-interest. *Investigation of firms* means finding out as much as possible about them so you can effectively promote your skills.

Self-analysis

As a student, your strongest selling point is your education; therefore, you should devote part of your letter and data sheet to describing it in terms of the potential employer's advantage. Many applicants have essentially the same educational background, however, so you must attempt to think of other items that distinguish you from your competitors. The following is a list of strengths that many employers view favorably.

1. Work experience related to your field: If you have been fortunate enough to have a part-time or full-time job in your technical area, emphasize this in your letter, perhaps devoting an entire paragraph to describing the work.
2. *Any* work experience: Some values have perhaps changed in this world, but employers still want someone who will give them a good day's work. Your experience may have consisted of frying hamburgers, but if you were a dependable employee who fried good hamburgers, potential employers will view this as evidence that you will do a good job for them. Mention this type of work in your data sheet.
3. College expenses earned: If you have paid for part or all of your college education, state the percentage in the education section of your data sheet.
4. Minor degree or sequence of useful courses: If you have earned a minor degree in an area that applies to the type of work you will be doing, mention this either in your letter or in your data sheet. Also mention sequences of courses beyond your school's basic requirements that might apply to your work. A sequence of three or more courses in management, writing, psychology, communication, or other areas might have given you knowledge or skills that your competitors do not possess.
5. Special projects in your major field: Toward the end of many 4-year

technical programs, students have an opportunity to work on a special project, sometimes called a senior project. Such projects permit students to investigate a particular area of their fields thoroughly, and to become more knowledgeable about that area than other students are. You should mention a project of this type in your data sheet or letter, especially if you have gained knowledge that companies might find useful.

6. Intent to continue your education: Most employers view favorably an employee's desire to upgrade his or her skills. If you intend to pursue a further degree, consider stating this in your data sheet or during your interviews with companies.

7. Grades: If your grade-point average is particularly good, state it in your data sheet; if your average is not particularly good, do not mention it anywhere.

8. Military experience: Applicants with military experience are generally viewed favorably by companies, so mention this in the experience section of your data sheet. If, while in the military, you attended a school that applies to your major field, identify the school in the education part of the data sheet.

9. Second language: Many of today's firms do business internationally, so if you have learned a second language at home or in school and believe it might be useful for a certain company, mention it in your data sheet in the education or personal-details section, whichever applies.

10. Membership in organizations: You should mention information about student or professional organizations in the personal-details section of your data sheet. If you have held an office in an organization or association, mention this also because it suggests that you have leadership ability and are able to work well with other people.

Investigation of Firms

After deciding what skills to emphasize, you can promote them much more effectively if you are aware of their market. You cannot use the you approach to its full potential unless you know how your skills might benefit each employer. Also, if your letter shows that you have taken the trouble to investigate a company, it will immediately distinguish you from your competition; most applicants know only that a job is available, and many do not even know that. Through investigation, you might also be able to find out which firms offer opportunities for advancement and which ones do not.

There are many sources of information about firms: school placement offices, where company literature and the *College Placement Annual* are available; career booklets such as *Careers in Technology;* professional journals; and professors. You should begin the investigative process long before you write application letters. After years of preparing for employment, do not sell yourself short by failing to examine the market thoroughly.

The remainder of this chapter is divided into sections aimed at two kinds of job applicants. The first section is aimed at students who have very little experience in industry and who will therefore emphasize their education when applying for a job. The latter part of the chapter provides help for students with full-time jobs in industry who will probably emphasize their experience to prospective companies. The two kinds of applicants will write different kinds of letters and data sheets; however, all students should read the entire chapter closely because much of the information applies to all applicants.

Letter Emphasizing Education

The application letter written by students with limited experience in industry contains three or four paragraphs. It is long enough to emphasize your main qualifications but short enough to invite reading. Its only function is to get an interview at which you can elaborate on your qualifications. It fulfills this objective by gaining the reader's attention, stating your main qualifications, and asking the reader for action—an interview.

ATTENTION

Although an application letter is essentially a sales letter, the indirect approach explained in Chapter 16 is not necessary. You can safely assume that your reader will be receptive to a brief, straightforward letter. Therefore, begin with a direct, positive paragraph that immediately specifies the job you are pursuing and summarizes your main qualification, your education. If you are not sure that an opening exists, apply for a particular kind of work that the firm is engaged in.

The "name beginning" provides a convenient opening sentence for the letter. It identifies the person, perhaps a school placement officer or a business associate of the firm, who suggested that an application letter be written. For the name beginning to work, the reader must be familiar with the name, and you must have received permission to use it.

If you do not use the name beginning, start with a sentence asking that

your qualifications for a certain job be considered. Conclude the first paragraph with a sentence that identifies education as your main qualification for the job. When you mention your education, inject the you approach, which is exemplified in the following samples of opening paragraphs:

> I would like you to consider my qualifications for the position of technician in your Chicago office. I understand that you are seeking someone to test radio equipment, and I believe that my two years of electrical engineering technology at Parker University have prepared me for this work.

> Professor John V. Elton, head of the Department of Mechanical Engineering at Parker University, has informed me of your opening in the area of fluid dynamics. I have specialized in fluid dynamics at Parker and am eager to learn more about this opportunity and your company.

> Has your Sterling Oaks Project opened any summer jobs in your drafting department? If so, I would like to show you what I can do, and you can help me determine what courses would be beneficial during my junior year in Parker University's School of Engineering.

AMPLIFICATION

In the second section of an application letter, elaborate on the major qualification you identified in the opening paragraph, your education. Start by naming the degree you will receive and when you will receive it. Then, mention areas of proficiency in your field that apply to the position you are seeking; this is preferable to naming courses in the letter because you will have an opportunity to do that in the data sheet. The sample paragraphs that follow show how to emphasize areas of proficiency.

The amplification section is your opportunity to point out particular strengths, so consider the 10 possibilities listed earlier in this chapter. Only the most pertinent information belongs in your letter, but a strength such as working experience in your field might well justify a second paragraph of amplification.

The end of the amplification section is the logical place to inform your reader that additional details can be found on the data sheet. Instead of just saying that a data sheet is enclosed, take advantage of this sentence to mention something not previously covered. For example, if the amplification section has been limited to your education, you might refer your reader to the data sheet for details about your education *and experience.*

I will receive an Associate Degree in Electrical Technology this June. My studies have centered on radio and audio equipment and have given me knowledge of troubleshooting techniques and circuit-analysis procedures. I have also gained experience with measurement equipment such as meters and oscilloscopes. The enclosed data sheet provides details about my education and work experience.

I will receive my Bachelor of Science Degree in Mechanical Engineering this June, and my program of study has emphasized areas that would prove helpful in your expanding Dynamics Division. In addition to specializing in fluid dynamics, I have gained knowledge of gas dynamics and thermodynamics. My studies have also included instruction in theoretical and applied instrumentation. The enclosed data sheet names specific courses I have taken and companies I have worked for while in school.

ACTION

The function of an application letter is not to secure a job, but to get an interview at which the job and your qualifications can be discussed in detail. Therefore, your final paragraph asks for an interview at the reader's convenience. If there are periods when you will not be available, give the reader this information.

Many firms are reluctant to invite a student to come 500 or 1000 miles for an initial interview. To schedule an interview with a distant firm, you can either suggest a meeting with a local representative, or mention that you will be visiting the firm's area during a certain period, perhaps a school vacation, and would appreciate an interview then. The firm's response will indicate whether traveling expenses will be paid, but you should be prepared to pay your own expenses if you suggest this type of arrangement. The following are examples of closing paragraphs:

May I have a personal interview with you to discuss my qualifications further? I can be reached at the telephone number or address at the top of my data sheet.

I will be visiting the Akron area during spring vacation, March 5-14, and I would appreciate having an interview with you. I can be reached at the address or telephone number listed on my personal record.

May I show you samples of my work? I am available for an interview at your convenience.

A complete application letter for a student emphasizing education rather than experience is shown in Fig. 17.1. The letter uses the semiblock format described in Chapter 15.

Data Sheet Emphasizing Education

A data sheet complements a concise letter by providing a detailed history of the applicant. In contrast to the letter, which is aimed at individual firms whenever possible, the data sheet usually remains the same. Photocopies are permissible if numerous applications are being sent. Readers prefer data sheets made up of concise, visually attractive data that allow quick assimilation. Needless to say, the data sheet must be complete. A year unaccounted for gives a negative impression.

As Fig. 17.2 shows, the top of the data sheet contains your name, address, and telephone number, along with the date. The rest of the data fall under four headings in the order of their importance: education, experience, personal details, and references.

EDUCATION

In the education section, list information in reverse chronology to accentuate your most recent achievements. Start this section by naming your college, degree, and date of graduation. Then list advanced courses that especially support statements made in your letter. You may include other information in the education section, such as grade-point average or percentage of college expenses earned, so review the 10 possibilities presented earlier in this chapter and try to think of additional selling points. At the end of the section, name your high school and date of graduation but do not elaborate unless some of your high school courses are applicable to the job you are seeking.

EXPERIENCE

The experience section is also arranged in reverse chronology. It includes dates, names and addresses of employers, and job titles. Emphasize any work that has included supervision of others, any job responsibilities related to the type of work you are seeking, and any promotions you have received. In either your experience or references section, consider providing the names of supervisors who will vouch for your performance at each company; ask for permission to use their names. List summer jobs as well as part-time jobs held for 6 months or more.

344 E. Azalea Ave.
Highland, IN 46322
March 1, 1977

Mr. William L. Smathers
Personnel Director
Mainland Manufacturing Co.
3618 Allen St.
East Chicago, IN 46312

Dear Mr. Smathers:

I would like you to consider my qualifications for the position of Junior Programmer in your Data Processing Department. I understand that you are looking for someone with a strong background in commercial programming languages, and my 2 years at Parker Community College have given me experience in this area.

I will receive my Associate Degree in Computer Programming this June. My courses in the computer field include 18 hours in the areas of commercial systems, computer systems analysis, and systems programming. While at Parker, I have become proficient with many types of unit-record equipment and a variety of digital computers. I have also gained extensive experience with the IBM 370 computer system, which I understand your company uses. The enclosed data sheet provides details about my education and work experience.

May I discuss this position and my qualifications with you at an interview? My phone number and address are at the top of the data sheet, and I am available at your convenience.

Yours sincerely,

Gail M. Smith

Gail M. Smith

Fig. 17.1   Application letter emphasizing education

GAIL M. SMITH
344 E. Azalea Ave.

319-924-1141   Highland, IN 46322   March 1, 1977

EDUCATION

1975-1977   Parker Community College, Hammond, IN
            I will receive my Associate Degree in Computer
            Programming in June 1977. My grade-point average
            is 5.18 of 6.00.

            Major Courses: RPG Programming
                            FORTRAN Programming
                            BAL Programming
                            COBOL Programming

            Background Helpful for This Position:
                        Operating Systems on IBM 370
                        Using Terminals with CDC 6500

1971-1975   Highland High School, Highland, IN
            Graduate, June 1975

WORK EXPERIENCE

1975-
Present   Full-Time During Summer, Part-Time During School
                        Secretary
                        Michael Jones Realty, 4325 Broadway
                        Cedar Lake, IN 46409

1973-1974   Full-Time During Summers
                        Waitress and Cook
                        Pizza Hut, Inc., 715 N. Calumet St.
                        Highland, IN 46322

PERSONAL DETAILS

Age: 20   Activities: President, Parker
Health: Excellent   Computer
Hobbies: Traveling, Drama   Club
                                        Treasurer, Highland
                                        Summer
                                        Theater

REFERENCES

Prof. James L. Lee   Ms. Susan Carlson   Mr. Michael Jones
Computer Department   Manager   President
Parker College   Pizza Hut, Inc.   Michael Jones Realty
Hammond, IN 46323   715 N. Calumet St.   4325 Broadway
                       Highland, IN 46322   Cedar Lake, IN 46409

Fig. 17.2   Data sheet emphasizing education

PERSONAL DETAILS

The personal-details section was once placed at the top of the data sheet, but wisdom and legislation such as the Civil Rights Act of 1964, the Age Discrimination Act of 1967, and the Equal Employment Opportunity Act of 1972 have relegated such details to a position near the bottom. Details about education and work experience tell employers more about an applicant's potential on the job than personal information regarding sex, age, religion, and marital status.

Ultimately, only you can decide what personal details to include in this section of the data sheet. Many applicants today choose to include only age, health, membership in student and professional organizations (as well as any offices held in them), and hobbies. It is not wise to attach a photograph to your data sheet because companies want to avoid the appearance of discrimination; some firms remove photographs, and others simply throw away the entire data sheet.

REFERENCES

In your references section, list the names, titles, and addresses of three or four people who will testify to your qualifications. When you ask for permission to use people's names, give them every opportunity to refuse your request; you do not want luke-warm references, and someone who refuses your request does you as much of a favor as someone who agrees to serve as a reference. Ideally, the list of references should include people who can vouch for your character, ability as a student, and performance as an employee.

Letter-Résumé Emphasizing Experience

The remainder of this chapter explains and exemplifies the application letter and data sheet, or résumé, typically written by people with several years of experience in industry. Although the letter and résumé are different from those described earlier, the self-analysis and investigation of prospective firms emphasized at the beginning of this chapter are important to the success of any job application.

LETTER

Applicants with experience typically write an extremely concise application letter that serves to transmit the résumé and to summarize the important information contained in the résumé. The letter consists of a paragraph for the first three items in the following list, and a one-sentence paragraph for the fourth item:

1. Identify your job objective.
2. Identify your major qualification, experience, for the position.
3. Summarize your experience in two or three sentences, and refer your reader to the résumé for detailed information.
4. Request an interview.

Figure 17.3 is a sample letter that goes along with the résumé in Fig. 17.4.

RÉSUMÉ

In a résumé, most of the information is grouped into sections similar to those in the data sheet. Employment history generally appears before education, however, and an objective section precedes both of them. As you read about each section, examine the sample résumé in Fig. 17.4.

Objective After placing your name, address, telephone number, and the date at the top of a résumé, you are expected to state your job objective in one or two sentences. Stating your objective is not as easy as it might seem because you have to accomplish several objectives at once. First, you should consider using the you approach by relating your objective to the company's objectives of increasing sales, production, quality of products, or efficiency. Second, you should attempt to indicate your immediate objective without excluding longer-term objectives.

You might find it helpful to think in terms of what you want to be doing 5 years from now, and to state your objective in a way that is both realistic and positive. Be careful of naming job titles because they frequently mean different kinds of work at different companies.

Employment History In the employment-history section, list in reverse chronology the firms for which you have worked. Provide details about your job responsibilities at companies during the past 10 years. Emphasize experience related to the position you are seeking, and give prominence to promotions you have received. Generally, do not go into detail about jobs held over 10 years ago; simply name the companies and positions held.

Education The education section of most résumés is basically the same as the education section of a data sheet. You should start by naming your

7094 Schneider Avenue
Hammond, IN 46323
March 3, 1977

Ms. Carol G. Stacy
Personnel Director
Jackson Engineering
1653 Lake Street
Louisville, KY 40214

Dear Ms. Stacy:

I would like you to consider my qualifications for the
position of Production Manager with your company. I have
been employed by Lane Company, Arlington Heights, IL, for 5
years. As Assistant Production Manager during the past 2
years, I have been responsible for modernizing Lane's assembly line, and the result has been a 35% increase in output.
Details about my employment and Bachelor's Degree in Industrial Management are provided in the enclosed résumé.

May I have an interview with you to discuss your position and my qualifications? My address and phone number are
listed on the résumé.

Sincerely yours,

George R. Kleinfelt

George R. Kleinfelt

Fig. 17.3   Application letter emphasizing experience

Résumé
of
GEORGE R. KLEINFELT
7094 Schneider Avenue
219–924–7265   Hammond, IN 46323   March 3, 1977

OBJECTIVE

I am seeking a position as Production Manager at a growing
firm that wants to modernize its facilities and methods to
increase production.

EMPLOYMENT HISTORY

1972–Present   Lane Company, Arlington Heights, IL, a multi-
                 plant manufacturer of automobile body parts.

        1975 – Promoted to Assistant Production Manager

                 Responsible for modernizing assembly–line
                 equipment and methods. Increased output by 35%
                 in 2 years.

        1972 – Employed as Maintenance Supervisor

                 Responsible for scheduling and supervising re-
                 pairs of assembly–line equipment

1970–1972   Myers Manufacturing Company, Crystal Lake, IL,
                 a manufacturer of hydraulic equipment.

                 Supervisor of Stamping Department

                 Responsible for direct supervision of 40
                 skilled–trade workers in metal–stamping de-
                 partment.

EDUCATION

1973–Present   Parker University, Hammond, IN. Will receive
                 Bachelor of Science Degree in Industrial Man-
                 agement in June 1977.

                 Major Courses: Production Cost Analysis
                                 Production Planning and Control
                                 Human Relations in Industry

Résumé of GEORGE R. KLEINFELT, page 2

1968–1970   Prairie Community College, Griffith, IL
Received Associate Degree in Mechanical
Technology in June 1970.

Major Courses: Metallurgy
                  Materials and Processes
                  Hydraulic and Pneumatic Sys-
                  tems
                  Production Problems

1962–1966   Bloom Township High School, Lancaster, IL
Graduated in June 1966.

MILITARY SERVICE

1966–1968   Served in United States Marine Corps for 19
months, 13 of which were in Vietnam. Honor-
ably discharged as sergeant in August 1968.

PERSONAL DETAILS

Age--29   Health--Excellent
Married--No Children   Hobby--Restoring Automobiles

AFFILIATIONS

American Society of Professional Foremen
Society for Technical Communication

REFERENCES

Available on Request

Fig. 17.4 Résumé emphasizing experience

college, degree, and date of graduation. Then, list courses that you believe would be useful, either immediately or in the future, at the company to which you are writing.

Education is generally placed after employment history in a résumé. Some résumé writers, however, have pursued a degree in order to change the direction of their careers, perhaps from a technical field to an area of management. If this is your situation, you probably should place education before employment history in the résumé and give particular emphasis to education in your letter.

Personal Details   Résumé writers frequently use more than one heading for personal information. For example, Fig. 17.4 has headings for personal details, professional affiliations, and military service; other résumés sometimes have a heading for community service. In short, a résumé writer has the freedom to include any heading that he or she can justify. Whatever the headings, a résumé typically mentions age, health, membership in professional or community organizations, and hobbies. Applicants must decide for themselves whether to include such personal information as marital status.

References   A résumé writer has the option of listing references or stating that names are available on request. To list references, some applicants would have to ask permission of people where they presently work; résumé readers understand why such applicants prefer not to publicize the fact that they are applying for work at another company.

The Interview

At an interview, you should be prepared to answer questions about your qualifications and career objectives, but there is no reason for the interview to be one-sided. Interviewers are generally quite willing to discuss how well your goals fit into the firm's future. In addition to specific questions about the position you are seeking, you can inquire about opportunities for advancement, the advisability of continuing your education, or any other matter that concerns you. If the interviewer does not mention salary, you may ask about it; however, most applicants find that firms offer approximately the same salaries and ultimately evaluate them according to other criteria. In short, if you have sufficiently researched potential firms, you know enough to ask the right questions at interviews. While showing your intelligence and interest, you gain the information necessary to make a wise decision.

Jobs are seldom offered at initial interviews, but you should not allow an interview to end until you receive some indication of your standing. A firm can reasonably be expected to set a date for either a second interview or a decision about your employment. If offered a job, you have a similar obligation; you are not expected to make an instant decision, but a date should be set for your response.

Follow-up Letter

After an interview with a particularly appealing firm, you can take one more step to distinguish yourself from the competition. Very few applicants write follow-up letters, although it takes only a few minutes to thank the interviewer and express your continued interest in the job:

```
     Thank you for our interview yesterday.  Our discussion
of Cranston's growing fluid dynamics' division was very in-
formative, and I am eager to make a contribution to it.

     I am looking forward to hearing from you.
```

Appendix

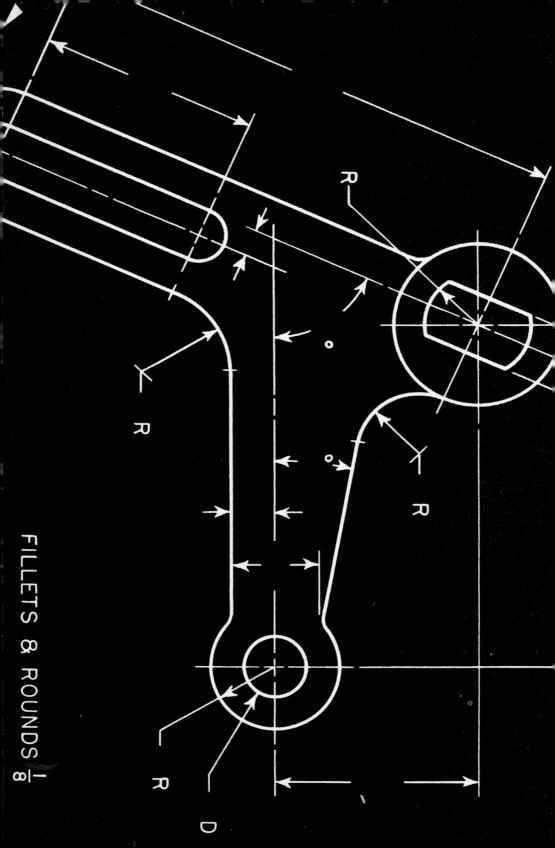

FILLETS & ROUNDS $\frac{1}{8}$

Common Technical Writing Errors

THIS appendix is essentially a reference or troubleshooting section. It attacks technical writers' most common stylistic and grammatical problems, and it emphasizes the objectivity, conciseness, and clarity required in reports. You may, however, want to purchase an English handbook for a more comprehensive coverage of grammar.

Person

Although the personal pronouns *I* and *we* invariably increase a report's readability, these words are generally not used in technical writing. Report readers expect objective thinking and writing, and they want you to reinforce your role as a detached, scientific observer by avoiding references to yourself. With rare exceptions, you will be expected to say such things as "The investigation showed that the transistors were faulty," rather than "My investigation showed. . . ." Also, avoid the word *you* in reports; nobody knows who "you" is.

Voice

Verbs are either intransitive or transitive. As the word *intransitive* implies, intransitive verbs do not transfer action:

Intransitive
The senior engineer <u>arrived</u>.
(The verb *arrived* has no direct object; it completes the action.)

Intransitive
A servomechanism <u>is</u> a feedback system.
(The verb *is* has no direct object; it simply links *feedback system* to *servomechanism*.

Transitive verbs, on the other hand, have a receiver of the action. Transitive verbs are either passive or active:

Passive
The electrode <u>is surrounded</u> by an insulator.
(The subject *electrode* receives the action of the verb.)

Active

An insulator surrounds the electrode.

(The subject *insulator* performs an action. The change from passive to active voice is accomplished by making *insulator* the subject of the sentence rather than the object of the preposition *by*; the previous subject *electrode* becomes the direct object in the active sentence.)

Use the active voice wherever possible in reports. It allows you to write more natural, direct, and concise sentences, and it places the emphasis on the performer of the action rather than on the receiver. When *you* are the performer of the action, however, report writing's conventional avoidance of personal pronouns usually forces you to say "It is recommended" rather than "I recommend."

Tense

Except for references to past and future events, which demand shifts in tense, use the present tense in technical writing.

Unnecessary Shift

Studies <u>show</u> that the procedure no longer <u>works</u>. It <u>was</u> slower and more time-consuming than the proposed procedure.

Uniform Tense

Studies <u>show</u> that the procedure no longer <u>works</u>. It <u>is</u> slower and more time-consuming than the proposed procedure.

Diction

Part of your job as a technical writer is to make complex ideas understandable. To help accomplish this, use concrete words and avoid redundancies and wordiness.

CONCRETE WORDS

Given a choice between abstract, highly technical words and simple, concrete, specific words conveying the same meaning, always choose the simple ones; if you must use a potentially confusing word, define it. Jargon, the

specialized language used and understood within a department or engineering discipline, should also be avoided in reports that travel beyond the department.

General terms, like *excessive, numerous,* and *frequently,* and words of judgment, like *effective, mediocre,* and *significant,* should be avoided or immediately clarified. They tell the reader very little in comparison with less ambiguous, more specific words.

REDUNDANCY

Redundancy is the repetition of an idea in different words. Here are some common redundancies and their corrections:

employed the use of	used
basic fundamentals	fundamentals
completely eliminate	eliminate
alternative choices	alternatives
actual experience	experience
the reason is because	the reason is that
connected together	connected
final result	result
prove conclusively	prove
rectangular in shape	rectangular

WORDINESS

From the report reader's standpoint, a good report says what needs to be said and gets it over with. This does not mean that all your reports should be short; however, no report should be longer than it has to be. Words that do not move the report forward are deadwood, needless obstacles to communication. At best they waste your reader's time, and at worst they blur your meaning. Many reports are cluttered by wordy phrases like "at this point in time" (now), and "come in contact with" (contact).

Wordy
Physical equilibrium is a facet of nature that is present in the total environment.

Revised
Physical equilibrium exists everywhere.

Wordy
Experience is relevant to the question as to whether or not a hypothesis is true.

Revised
Experience indicates whether a hypothesis is true.

Wordy
When a body is in static equilibrium, it shows a lack of movement.

Revised
A body in static equilibrium does not move.

Revise the following sentences, removing unnecessary words.

1. Studies of the two units are needed to be conducted to determine which of them would be more economical to purchase.

2. A complete orientation time of about 30 hours is suggested as a sufficient time period for learning the basic and essential operations of the system.

3. The location of the data collector will be in the presently vacated storage room.

4. From answers on the questionnaires it was found that the average amount of use that these units will receive is 100 hours per month.

5. A commonly known fact is that of expansion and contraction of materials, steel being no exception.

6. The alternatives must be examined carefully so that the money spent will be worth the investment.

7. Inefficiency was said to be the most important factor in the decision to hire a new supervisor.

8. The manufacturing company of this system strongly urges cleaning of the system so as to keep intricate parts from every once in a while becoming jammed.

9. Several housing projects are found 2 miles south of Springfield and can be seen to be setting a trend of moving parallel with the highway.

10. Several manufacturers have developed new bumpers to cushion the impact of a car in a collision with another car or in striking an object.

Sentences

A sentence is a group of words containing a subject and a verb and expressing a complete thought. Sentences may be composed of smaller subject-verb units called clauses.

MAIN AND SUBORDINATE CLAUSES

Subject-verb units that express complete thoughts are called main, or independent, clauses. If the words forming an independent clause were removed from a sentence, they would form a complete thought by themselves. The two independent clauses in the following sentence are underlined: <u>A 3000-megawatt nuclear plant requires 1200 acres</u>, but <u>smaller plants require less land</u>.

Subordinate, or dependent, clauses contain subject-verb units that cannot stand alone; to make sense, they must be linked to independent clauses. The subordinate clause is underlined in the following sentence: Water from nuclear plants must be cooled <u>because it endangers aquatic life</u>. Subordinate clauses serve as nouns, adjectives, or adverbs.

Noun Clause
The experiment showed <u>that the new material was flawless</u>.
(The clause serves as a direct object of the verb *showed*.)

Adjective Clause
The information <u>that the researchers gathered</u> did not prove anything.
(The clause serves as an adjective modifying the noun *information*.)

Adverb Clause
Sound waves are radiated <u>when the voice-coil assembly is vibrating</u>.
(The clause serves as an adverb modifying the words *are radiated*.)

PHRASES

A phrase is a group of related words not containing a subject and a verb. Phrases may serve as nouns, adjectives, or adverbs, and are introduced by prepositions or verbals.

Prepositional Phrases
In prepositional phrases, the preposition connects a noun or pronoun to the rest of the sentence. Prepositions include the words *with, without, at, on, from, about, by, for, during, in, of, through, to, until,* and *under.* Prepositional phrases serve as either adverbs or adjectives.

The current was connected to the voltmeter.
(The phrase serves as an adverb modifying *was connected.*)

The molecules of the metal were magnetized.
(The phrase serves as an adjective modifying *molecules.*)

Verbal Phrases
Words derived from verbs and used as nouns, adjectives, or adverbs are called verbals. Gerunds, participles, and infinitives generally join other words to form verbal phrases.
Gerunds always end in *-ing* and serve as nouns:

The system began injecting fuel.
(The phrase serves as the direct object of the verb *began.*)

Participles always serve as adjectives. Present participles end in *-ing* and past participles usually end in *-ed, -en,* or *-t:*

Nothing stopped the workers installing the computer.
(The phrase serves as an adjective modifying *workers.*)

Built in Texas, prepared by the Wood brothers, and driven by A. J. Foyt, the Coyote-Ford won the race.
(The phrases serve as adjectives modifying *Coyote-Ford.*)

Infinitives are the *to* forms of verbs, although the *to* is sometimes omitted. Infinitives serve as nouns, adjectives, and adverbs:

To begin the project involved many risks.
(The noun phrase serves as the sentence's subject.)

It was a difficult project <u>to begin</u>.
(The adjective phrase modifies the noun *project*.)

The department was eager <u>to begin</u> the project.
(The adverb phrase modifies the verb *was eager*.)

RESTRICTIVE AND NONRESTRICTIVE MODIFIERS

Modifiers that restrict the meaning of a noun are said to be restrictive. *Restrictive modifiers* provide information important for proper interpretation of the sentence. They are not set off by commas.

Nonrestrictive modifiers contain parenthetical information that does not change the meaning of the sentence. Therefore, nonrestrictive modifiers are set off from the rest of the sentence by commas.

Nonrestrictive
He paid the taxes, <u>which he considered fair</u>.

Restrictive
He paid the taxes <u>that he considered fair</u>.

Nonrestrictive
The tests, <u>which proved nothing</u>, will be skipped next time.

Restrictive
The tests <u>that proved nothing</u> will be skipped next time.

MISPLACED AND DANGLING MODIFIERS

Sentences become confusing when modifiers do not point directly to the words they modify. Misplaced and dangling modifiers often produce absurd sentences; worse yet, they occasionally result in sentences that make sense, causing the reader to misinterpret your meaning. Modifiers must be placed in a position that clarifies their relationship to the rest of the sentence.

Misplaced
<u>Before leaving the company</u>, an inspector examines the products carefully.

Correction
An inspector examines the products carefully <u>before they leave the company</u>.

Misplaced
Thermal pollution is worse at nuclear facilities, <u>which must be reduced</u>.

Correction
Thermal pollution, <u>which must be reduced</u>, is worse at nuclear facilities.

PARALLEL STRUCTURE

Sentences are like equations. The structure of a sentence helps clarify your meaning by reinforcing the relationship of your ideas. Elements of a sentence that are connected by coordinating conjunctions (*and*, *but*, *or*, *nor*, *for*, *yet*, *so*) carry the same weight and must be grammatically similar, or parallel. Items in a series of words, phrases, or subordinate clauses have the same value, and their grammatical structure must indicate their equality. If coordinate elements in a sentence do not have a parallel pattern, the sentence becomes awkward and potentially confusing.

Faulty
Management guarantees <u>that the old system will be replaced</u> and <u>to consider the new proposal</u>.
(The subordinate clause and infinitive phrase do not have parallel structure.)

Parallel
Management guarantees <u>that the old system will be replaced</u> and <u>that the new proposal will be considered</u>.

Faulty
Use the cylinder with <u>a diameter of $3^3/_{16}$ inches</u> and <u>1½ inches high</u>.
(To be parallel, the two items in this series should be objects of the preposition *with*).

Parallel
Use the cylinder with <u>a diameter of $3^3/_{16}$ inches</u> and <u>a height of 1½ inches</u>.
(Parallelism can also be achieved by making both items objects of the verb *has*.)

Parallel
Use the cylinder that has <u>a diameter of $3^3/_{16}$ inches</u> and <u>a height of 1½ inches</u>.

Faulty

A successful firm is capable of <u>manufacturing a product, marketing it,</u> and <u>make a profit.</u>
(For this series to be parallel, all three objects of the preposition *of* must be participles.)

Parallel

A successful firm is capable of <u>manufacturing a product, marketing it,</u> and <u>making a profit.</u>

Revise the following sentences, making their coordinate elements parallel.

1. A proposal for the use of the land and how to finance the project has been submitted.

2. The manual contains directions for excavating earth, laying asphalt, and the installation of synthetic turf.

3. Steps must be taken to remove phosphates from waste material, devising a system for early detection of oil spills, and creation of laws to prohibit the dumping of industrial waste.

4. The most frequent errors are those caused by improper calibration or that are the result of careless operation.

5. A monthly charge of $400 is for computer time, printed output, and to maintain the system.

6. The measuring device's advantages are its reduction of errors, requiring less maintenance, and provides greater accuracy.

7. The shaft, with a diameter of ⅜ inch and 1¼ inches long, is grooved toward the end.

8. A surveying instrument should be purchased not only because it reduces human errors but also resulting in saved time.

9. A shaft extends from the socket holder, then going through the ratcheting gear, and to a push button.

10. The Wankel, which is a thoroughly tested engine and used by Japanese auto manufacturers, will cost less when full production begins.

SUBJECT-VERB AGREEMENT

Both the subject and the verb of a sentence must be singular or plural. Almost all problems with agreement are caused by failure to identify the subject correctly.

Faulty
The stockpile of chemicals are located in an uninhabited area.
(The subject *stockpile* is singular; *chemicals* is the object of a preposition.)

Correction
The stockpile of chemicals is located in an uninhabited area.

Faulty
The distances that E and W move around the fulcrum determines the amount of effort needed.

Correction
The distances that E and W move around the fulcrum determine the amount of effort needed.

Faulty
The committee are investigating the sales report.
(When a *collective noun* refers to a group as a unit, the verb must be singular. Other collective nouns are *management, union, team, audience,* and *jury*.)

Correction
The committee is investigating the sales report.

Faulty
Each of the steelworkers are highly skilled.
(Indefinite pronouns, such as *each, everyone, either, neither, anyone,* and *everybody*, take a singular verb.)

Correction
Each of the steelworkers is highly skilled.

Faulty
Neither the foreman nor the laborers wants a strike.
(When compound subjects are connected by *or* or *nor*, the verb must agree with the nearer noun.)

Correction
Neither the foreman nor the laborers want a strike.

The word *data* causes some confusion for technical writers. Traditionally, the word has been considered plural, but the trend is toward viewing it as singular, in the sense of information. Today, you may use either a singular or plural verb with *data*, as long as you are consistent throughout the report.

Singular
The data is. . . . It. . . . This data. . . .

Plural
The data are. . . . They. . . . These data. . . .

Select the correct verbs in the following sentences

1. Cable for the zoom, pan and tilt mechanisms (cost, costs) $60 per 1000 feet.

2. The piston engine now used by all U.S. manufacturers has been developed to the point where further reductions in pollution (seem, seems) unlikely in the near future.

3. Following the first section of the report (is, are) descriptions of the system and its installation.

4. None of the employees (is, are) going to like the revised vacation schedule.

5. The function of the cables (is, are) to provide voltage required by the zoom lens.

6. These facts about water pollution in Lake Michigan (was, were) gathered from an article about the problem.

7. For some operations with a slide rule, the choice of scale combinations (make, makes) no difference.

8. The head is made of chrome-plated, drop-forged steel that (encase, encases) the ratchet and socket-releasing mechanisms.

9. The cables that connect the camera to a monitor (carry, carries) the video signal.

10. Neither the surveyor nor the engineer (perform, performs) routine operations in the field.

PRONOUN REFERENCE

A pronoun must refer directly to the noun it stands for, its antecedent. Pronouns commonly used in technical writing include *it, they, who, which, this,* and *that.* Never use *they* as an indefinite pronoun; when you write "They say that," make sure your reader knows who "they" are. *It* may be used sparingly as an indefinite pronoun ("It is obvious that"), but overuse of the indefinite *it* leads to confusion.

As in subject-verb agreement, both a pronoun and its referent must be singular or plural. Collective nouns generally take the singular pronoun *it* rather than *they.*

Problems result when pronouns such as *they, this,* and *it* are used carelessly, forcing the reader to figure out their referents:

Vague
Research teams examined sites for a new chassis factory. <u>They</u> overlooked the lakefront.

Clear
Research teams examined sites overlooking the lakefront for a new chassis factory.

Vague
The consultants recommended a new method for selling aluminum. <u>This</u> is the company's best bet for the future.

Clear
The consultants recommended a new method for selling aluminum, the company's best bet for the future.

Vague
The Atomic Energy Commission determines the criteria for selecting a nuclear site. It includes a specified distance from high-population zones.

Clear
The Atomic Energy Commission, which determines the criteria for selecting a nuclear site, specifies that the site be a certain distance from high-population zones.

Revise the following sentences, making the pronoun references *unmistakable*.

1. The faster test drivers drove the Corvettes, the quieter they got.

2. Federal officials say that auto manufacturers must adhere to new laws restricting auto emissions. They brought them on themselves.

3. The company had high hopes for a new research program, but it encountered financial problems.

4. Asphalt will be spread over the field as soon as excavation is completed. This takes approximately 5 weeks.

5. He will explain how the new components can be adapted to the existing systems as well as their basic principles.

6. During the compression stroke, the piston moves up, compressing the mixture. Then the mixture ignites, and expanding gas forces it down.

7. The field can be covered during the football season, which requires approximately 45 minutes.

8. Several companies are competing with General Motors and Ford, so they must improve their research programs.

9. Compasses should be carried by all personnel in the swamp to avoid being lost.

10. When the supervisor read a report about personnel sleeping on the job, she had no alternative but to consider it carefully.

COMMAS, COLONS, AND SEMICOLONS

The function of punctuation is to help clarify the meaning of your sentences. A punctuation mark as seemingly unimportant as a comma can radically change your reader's interpretation of a sentence. This section explains the basic uses of the comma, colon, and semicolon.

The Comma
Use commas in the following ways:

1. To separate two main clauses connected by a coordinating conjunction (*and, but, or, nor, for, yet, so*). The comma may be omitted if the clauses are very short.

> *Two Main Clauses*
> Radar feeds the information back to the control station, and a new course is relayed to the missile.

2. To separate introductory subordinate clauses or phrases from the main clause.

> *Clause*
> When the target changes course, radar detects the change.

> *Phrase*
> Sensing a reduction in pressure, the regulator sends more gas into the pipeline.

3. To separate words, phrases, or clauses in a series.

> *Words*
> The recorder contains a noise-reduction unit, a frequency-equalizer unit, and a level-control unit.

> *Phrases*
> The skill is needed in industry, in education, and in government.

> *Clauses*
> Select equipment that has durability, that requires little maintenance, and that the company can afford.

4. To set off nonrestrictive appositives, phrases, and clauses. (Dashes and parentheses also serve this function. Parentheses may be used frequently, but dashes should be used sparingly.)

Appositive
Clem Stacy, the supervisor, fixed the generator.

Phrase
The engine, beginning its seventh year of service, finally needed maintenance.

Clause
The system, which is somewhat complicated, requires calibration at various intervals.

5. To separate coordinate but not cumulative adjectives.

Coordinate
He rejected the distorted, useless recordings.
(The adjectives are coordinate because they modify the noun independently. They could be reversed with no change in meaning: *useless, distorted recordings*.)

Cumulative
An acceptable frequency-response curve was achieved.
(The adjectives are not separated by commas because they modify intervening adjectives as well as the noun. Cumulative adjectives cannot be reversed without distorting the meaning: *frequency-response acceptable curve*.)

6. To set off conjunctive adverbs and transitional phrases.

Conjunctive Adverbs
The new regulations, however, caused a morale problem.

The crane was very expensive; however, it paid for itself in 18 months.

Therefore, the branch plant should not be built until 1976.

Transitional Phrases
On the other hand, the maintenance crew operated efficiently.

Performance on Mondays and Fridays, for example, is far below average.

The Colon
Use colons in the following ways:

1. To separate an independent clause from a list of supporting statements or examples.

A piston has the following strokes: intake, compression, power, and exhaust.

2. To separate two independent clauses when the second clause explains or amplifies the first.

The Wankel engine is potentially better than conventional engines for one major reason: it has 40% fewer parts.
(Do not capitalize the first word of the second independent clause unless you want to give the clause special emphasis.)

The Semicolon
Use semicolons in the following ways:

1. To separate independent clauses not connected by coordinating conjunctions (*and, but, or, nor, for, yet, so*).

The storm stopped the surveyors; in fact, it stopped all work at the site.

2. To separate independent clauses connected by coordinating conjunctions *only* if the clauses are long or have internal punctuation.

In a gunfire-control system, the target plane is moving, and the input information is variable, depending on the plane's speed and range; but radar, acting as the sensing device, feeds the input information to the control station.

3. To separate independent clauses when the second one begins with a conjunctive adverb (*therefore, however, also, besides, consequently, nevertheless, furthermore*).

The machine performs better than all the others; therefore, it should be purchased.

4. To separate items in a series if the items have internal punctuation.

Plants have been proposed for Kansas City, Missouri; Seattle, Washington; and Orlando, Florida.

Nonsentences

Comma splices, run-on sentences, and sentence fragments are among the most serious sentence problems. This section shows how to recognize them and correct them.

COMMA SPLICES

A comma splice occurs when two independent clauses are connected, or spliced, with only a comma. You can correct comma splices in four ways:

1. Replace the comma with a period to separate the two sentences.

Splice
Friction results from the movement of one material across another, uneven surfaces cause greater friction.

Correction
Friction results from the movement of one material across another. Uneven surfaces cause greater friction.

2. Replace the comma with a semicolon *only* if the sentences are very closely related.

Splice
A hypothesis is an assumption, therefore it must be tested.

Correction
A hypothesis is an assumption; therefore, it must be tested.
(The word *therefore* is a conjunctive adverb. When you use a conjunctive adverb to connect two sentences, always precede it with a semicolon and follow it with a comma. Other conjunctive adverbs are *however, also, besides, consequently, nevertheless,* and *furthermore.*)

3. Insert a coordinating conjunction after the comma, making a compound sentence.

Splice
The gas-air mixture burns in internal combustion engines, it does not explode.

Correction
The gas-air mixture burns in internal combustion engines, but it does not explode.
(The words *and, but, or, nor, for, yet,* and *so* are coordinating conjunctions.)

4. Subordinate one of the independent clauses by beginning it with a subordinating conjunction or a relative pronoun. Do not use this method for correcting a comma splice unless the clause *should* be given less emphasis.

Splice
The operation of a radar system includes three main sequences, they are the generation, transmission, and reception of a signal.

Correction
The operation of a radar system includes three main sequences, which are the generation, transmission, and reception of a signal.
(The words *which, that, who,* and *what* are relative pronouns. Frequently used subordinating conjunctions are *where, when, while, because, since, as, until, unless, although, if,* and *after.*)

Correct the following comma splices:

1. During the first year, the company saved $20,000, these savings resulted from better inventory control.

2. The condition, which steadily grows worse, will continue to deteriorate, action to alleviate the pollution problem, including the creation of new laws, is badly needed.

3. Several antennas have been considered, one of them, the "big wheel" antenna, satisfies both criteria.

4. The system operates simply and efficiently, also it keeps track of the current progress of the items.

5. When multiplying with the slide rule, the result is shown at the right index if the slide is moved to the left, however, the result is shown at the left index if the slide is moved to the right.

6. As the temperature fluctuates, so does the length of the tape, if this is not taken into consideration, precision suffers.

7. A carburetor's operation is based on the Venturi principle, that is, a gas or liquid flowing through a restriction will increase in speed and decrease in pressure.

8. Two valves control fluid flow, one releases pressure in the system and allows the ram to retract, the second controls the pistons.

9. When the iron becomes magnetized, it can attract other pieces of iron, when no current is flowing in the coiled wire, the iron is not magnetized.

10. Consequently, the area will need a new airport, in fact, construction should start immediately.

RUN-ON SENTENCES

Run-on, or fused, sentences look like comma splices without the comma. The independent clauses are run together with no punctuation between them. To eliminate fused sentences, use one of the four methods explained above: (1) place a period between the two clauses, (2) place a semicolon between them, (3) place a comma and a coordinating conjunction between them, or (4) place a relative pronoun or subordinating conjunction between them.
   Correct the following run-on sentences.

1. Jerry Chapetta's article is entitled "Great Lakes: Great Mess" it appears in the May 1968 issue of *Audubon*.

2. According to Chapetta, the chemical symbol for water in the Great Lakes should be H_2Os_{10} this symbol stands for 2 parts hydrogen, 1 part oxygen, and 10 parts stupidity.

3. During the installation period, business operations will not be disrupted switching over to the new system will take less than a day.

4. When the level of the capsule changes, the heater turns off the charge causes mercury to roll away from the contacts.

5. The function of the carburetor is to atomize the fuel and mix it with air flowing into the engine the carburetor must also meter the fuel to provide the proper fuel-air ratio.

6. An engine converts burning fuel into usable energy the energy is in the form of a rotating shaft.

7. The temperature in any part of the building can be controlled this is beneficial during periods of reduced occupancy.

8. Careful driving is important but the automobile should be designed for safety many people are more concerned about the auto's appearance.

9. The early air bags lacked a reliable sensor a sensing device had to be developed.

10. Lack of funds is the most important problem in pollution control for several years the government has promised assistance but very little has been delivered.

SENTENCE FRAGMENTS

Sentence fragments are incomplete thoughts that have been punctuated as complete sentences. Fragments are often subordinate clauses, prepositional phrases, and verbal phrases. As the following examples show, they must be connected to the preceding or following sentence to gain meaning.

1. Connect subordinate clauses to independent clauses.

Fragment
The company continues to lose money. Although production has increased. (The fragment is a subordinate clause beginning with a subordinating conjunction. Other subordinating conjunctions are *where, when, while, because, since, as, until, unless, if,* and *after.*)

Correction
The company continues to lose money although production has increased.

Fragment

The problem originates in the transformer. <u>Which does not provide enough
electric energy</u>.
(The fragment is a subordinate clause beginning with a relative pronoun.
Other relative pronouns are *who, that,* and *what*.)

Correction

The problem originates in the transformer, <u>which does not provide enough
electric energy</u>.

2. Connect prepositional phrases to independent clauses.

Fragment

Civil engineering requires many skills. <u>For example, drafting and surveying</u>.
(The fragment is a prepositional phrase. Other prepositions are *with, without, at, on, from, about, by, during, in, of, through, to, until,* and *under*.
The fragment can be converted into a subordinate clause, as in the first
example below, or made into a participial phrase.)

Correction

Civil engineering requires many skills, such as drafting and surveying.

Correction

Civil engineering requires many skills, including drafting and surveying.

3. Connect verbal phrases to independent clauses.

Fragment

Kinetics is a field of dynamics. <u>Consisting of all aspects of motion</u>.
(Verbal phrases, including this participial phrase modifying *field*, often
begin with *-ing* words. Such phrases must be linked to independent clauses.)

Correction

Kinetics is a field of dynamics consisting of all aspects of motion.

Fragment

The writer studied the statistics. <u>To insure his accuracy</u>.
(Infinitive phrases, including this one modifying *studied*, often begin with
to and a verb. They must be linked to independent clauses.)

Correction
The writer studied the statistics to insure his accuracy.

Correct the following sentence fragments.

1. Although these experiments are performed by experienced people at the laboratory.

2. Human beings, the cause of all pollution in Lake Michigan and the only hope for saving the lake.

3. When atoms absorb the spark's energy and give off light.

4. Material presented to emphasize the fate that seems to be awaiting the environment.

5. The offending companies, having been named in the report along with evidence of their guilt.

6. Five elements are checked immediately. The other four sent to the laboratory for testing.

7. That the high-voltage system, being more efficient, causes fewer technical problems than other systems.

8. An electronic system has the same basic problems as a mechanical system; speed and accuracy.

9. Because the manual method of testing requires at least 7 minutes regardless of the employee's skill.

10. Each year more than $10,000 can be saved if the company buys a new measuring system. No decrease in quality if the measurement is accurate.

Paragraphs

A paragraph consists of several sentences anchored by a topic sentence. The topic sentence expresses the paragraph's central idea, and the remaining sentences develop, explain, and support the central idea.

TOPIC SENTENCES

The central idea of each paragraph must be stated in a strong topic sentence. In technical writing, the topic sentence usually appears at the beginning of a paragraph, followed by details that support and clarify the central idea. This deductive structure gives your paragraphs the direct, straightforward style preferred by most report readers. Inductive paragraphs, which may occasionally be used, begin with the details and build up to a topic sentence at the end of the paragraph.

All deductive paragraphs follow a statement-support pattern. The topic sentence consists of a generalization that must be explained or illustrated with details, as in the following example:

> Uranium will continue to be the major fuel for nuclear reactors. The latest statistics show that enough uranium is already available to supply the needs of the United States beyond 1990. The supply of uranium will increase as technology develops new methods of locating, mining, and processing it. As the supply increases, uranium's cost should decrease, making it even more attractive as a fuel for nuclear reactors.

Two kinds of statement-support paragraphs, effect-and-cause and comparison-contrast, are common in technical reports. Whichever pattern you use, insure that your topic sentence clearly fixes the direction and boundary of the paragraph.

EFFECT-AND-CAUSE

Begin an effect-and-cause paragraph by stating an effect; then explain, or trace, its causes:

> A coal shortage is developing in the United States, not because of the lack of coal, but because of decreased coal mining. Previously, the coal industry was plagued with overproduction, but mechanization has greatly increased the cost of opening new mines and dampened the enthusiasm of investors. Labor problems and the necessity for better health and safety equipment have also discouraged investors. Recent restrictions on the sulfur content of coal have increased the shortage; the industry now has difficulty mining enough coal of the quality required by the government to supply growing needs.

COMPARISON-CONTRAST

In paragraphs involving alternatives, the comparison-contrast pattern allows you to present data for the alternatives and to discuss each alternative's advantages and disadvantages:

> Operation and maintenance costs are lower for nuclear plants than for coal-fired plants. Nuclear plants require no equipment for handling and burning coal, which results in fewer employees and smaller payrolls. A coal plant needs approximately 250 people, but a nuclear facility can operate efficiently with approximately 100 workers. Therefore, the cost of running a coal plant is about 0.24 mills per kilowatt-hour, versus 0.10 mills for a nuclear station.

TRANSITION

Transition between and within paragraphs causes no problems if you present ideas in a logical and orderly way. Synonyms and repetition of key words achieve transition as you progress from idea to idea ("The company won 15 contracts last year. . . . These victories. . . . Most of the contracts. . . ."). To clarify the relationship of your ideas further, use connective words and expressions:

for example	on the other hand
for instance	on the contrary
in fact	therefore
in other words	furthermore
in short	consequently
in addition	similarly
however	in summary
nevertheless	in conclusion

Capitalization

The conventional rules of capitalization apply to technical writing. Within each report, capitalize uniformly and remember that the trend in industry is away from overcapitalization. For example, say "The senior project engineer, Mr. Reynolds," rather than capitalizing each word of his title, and say "Tinius Olsen testing machine" rather than capitalizing the name of the product as well as the trademark.

Use capitalization to provide emphasis and to indicate levels of headings within reports. Titles of reports, titles of visuals, and first- and second-level headings can be entirely capitalized. Capitalize the first word and each succeeding word (except prepositions, conjunctions, and articles) of third- and fourth-level headings and labels on visuals. Also, capitalize references to visuals ("As Fig. 4 shows").

Abbreviations

Use abbreviations only for long words or combinations of words that must be used more than once in a report. For example, if words like *Fahrenheit, cubic inches, pounds per square inch,* or *British thermal units* must be used several times in a report, abbreviate them to save space. Below are eight rules for abbreviating.

1. If an abbreviation might confuse your reader, spell it out in parentheses the first time you use it.

2. Use small letters except for the abbreviations of proper nouns such as *British thermal units* (Btu).

3. Do not add *s* to form the plural of an abbreviation.

4. Do not use abbreviations for units of measurement preceded by approximations. You may say "15 psi" but not "several psi."

5. Do not abbreviate short words such as *acre, ton,* or *mile.*

6. Omit the period after an abbreviation unless the abbreviation might be confused with some other word.

7. Do not use periods within or after acronyms (NASA). Also, do not space between the letters.

8. Use abbreviations (and symbols) when necessary to save space on visuals, but define difficult ones in the legend, footnote, or text.

Possessives

The following are four basic rules for showing possession:

1. Add an apostrophe and an *s* to show the possession of nouns that do not already end with *s*.

    a corporation's profits
    a woman's career
    the women's caucus

2. Add only an apostrophe to plural nouns ending with *s* ("eleven corporations' profits").

3. For singular nouns of one syllable ending with *s* add an apostrophe and an *s* ("boss's orders"; "punch press's cost") but add only an apostrophe if the singular noun has two or more syllables ("Maria Williams' job").

4. Do not add an apostrophe to personal pronouns (*theirs, ours, its*). The only time *its* needs an apostrophe (*it's*) is when it's a contraction for the words *it is*.

Hyphens

Use hyphens to connect two or more modifiers that express a unified, or one-thought, idea ("high-frequency system"; "alternating-current motor"). If some of the following modifiers were not hyphenated, confusion would result:

    energy-producing cells   plunger-type device
    8-hour shifts   foreign-car buyers
    cement-like texture   trouble-free system
    A-frame construction

The following are rules for hyphenating:

1. Use hyphens to connect compound modifiers formed from a quantity and a unit of measurement ("a 3-inch beam," "a 3-inch-wide beam," "an 8-mile journey," "an 8-mile-long journey") unless the unit is expressed as a plural

("a beam 3 inches wide," "a beam 3 inches in width," "a journey of 8 miles").

2. Use hyphens to connect compound nouns used as units of measurement ("kilogram-meter," "kg-m").

3. Use hyphens to connect compound numbers from 21 through 99 when they are spelled out ("Thirty-five safety violations were reported.") and fractions when they are spelled out ("three-fourths"). Hyphenate complex fractions if the fraction cannot be typed in small numbers ("1-3/16 miles"). Do not hyphenate if the fraction can be typed in small numbers ("1½ hp").

4. Use suspended hyphens for a series of adjectives that you would ordinarily hyphenate ("10-, 20-, and 30-foot beams").

5. Do not hyphenate -*ly* adverb-adjective combinations ("recently altered system"). Hyphenate other adverb-adjective combinations when they precede nouns ("well-developed system") but not when they follow nouns ("system that was less developed").

Numbers

The following rules cover most situations, but when in doubt whether to use a figure or a word, remember that the trend in report writing is toward using figures:

1. For numbers accompanying units of measurement, use figures:

1 gram	0.452 minute
33⅓%	$6.95
8 yards	5 weeks

2. When not used with units of measurement, spell out numbers below 10, and use figures for 10 and above:

four cycles	10 machines
two-thirds of the union	1835 members

3. When numbers above and below 10 are in a series, use figures for all of them.

4. For compound-number adjectives, spell out the first one or the shorter one to avoid confusion ("75 twelve-volt batteries").

5. Spell out numbers that begin sentences.

6. Express plurals of figures by adding 's ("16's," "1970's").

7. Use figures to record numbers on dials, and so on. ("Readings of 7.0, 7.1, and 7.3 were taken.")

8. Combine figures and words for extremely large numbers ("2 million miles").

9. For decimal fractions of less than one, place a zero before the decimal point ("0.613"). When precision demands, place a zero after the decimal point ("0.6130").

10. Place the last two letters of the ordinal after fractions used as nouns ("1/50th of a second"), but not after fractions that modify nouns ("1/50 horsepower"). Spell out ordinals below 10 ("third mile," "ninth man"), but for 10 and above, use the number and the last two letters of the ordinal ("10th day," "21st year").

Symbols

Like abbreviations, symbols save space but cause confusion. Except for the symbols for *dollars* and *percent*, symbols should not be used within the paragraphs of a report. To save space in tables and drawings, however, use simple symbols like the ones for *feet, inches,* and *degrees*. Use highly technical symbols in the report's appendix if you are certain that they will be understood.

Index